D1381263

Kingdom of Lions

Kingdom of Lions

JONATHAN SCOTT

KYLE CATHIE LIMITED

For my wife Angela,
whose love and support contributed so much to this book

First published in Great Britain in 1992 by
Kyle Cathie Limited
20 Vauxhall Bridge Road, London SW1V 2SA

ISBN 1 85626 061 5 hardback

A CIP catalogue record for this book is available from the British Library.

Designed by David Fordham
Maps by Caroline Simpson
Photoset by Rowland Phototypesetting Limited,
Bury St Edmunds, Suffolk
Produced by Mandarin Offset Limited
Printed in Hong Kong

Contents

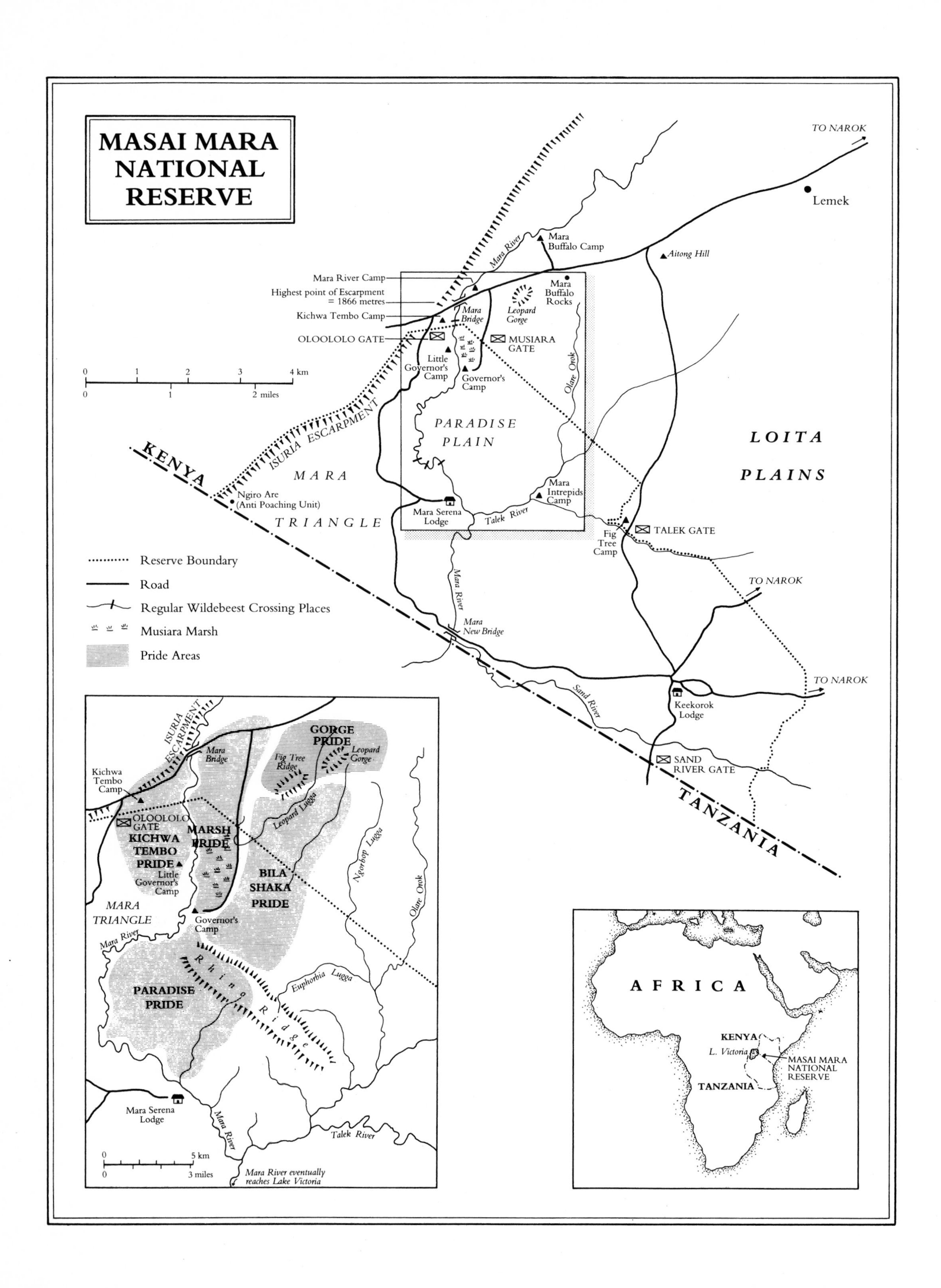

MASAI MARA NATIONAL RESERVE
TO NAROK
Lemek
Mara Buffalo Camp
Aitong Hill
Mara River
Mara River Camp
Highest point of Escarpment = 1866 metres
Kichwa Tembo Camp
OLOOLOLO GATE
Mara Bridge
Mara Buffalo Rocks
Leopard Gorge
MUSIARA GATE
Little Governor's Camp
Governor's Camp
Olare Orok
ISURIA ESCARPMENT
PARADISE PLAIN
LOITA PLAINS
KENYA
MARA TRIANGLE
Ngiro Are (Anti Poaching Unit)
Mara Serena Lodge
Talek River
Mara Intrepids Camp
Fig Tree Camp
TALEK GATE
Reserve Boundary
Road
Regular Wildebeest Crossing Places
Musiara Marsh
Pride Areas
Mara New Bridge
Sand River
Keekorok Lodge
SAND RIVER GATE
TANZANIA
0 1 2 3 4 km
0 1 2 miles
GORGE PRIDE
Fig Tree Ridge
Leopard Lugga
KICHWA TEMBO PRIDE
MARSH PRIDE
BILA SHAKA PRIDE
Ngorbop Lugga
PARADISE PRIDE
Rhino Ridge
Euphorbia Lugga
0 5 km
0 3 miles
Mara River eventually reaches Lake Victoria
AFRICA
KENYA
L. Victoria
TANZANIA
MASAI MARA NATIONAL RESERVE

Introduction

Exalted and denigrated, admired and despised, no animals have so aroused the emotions of man as have the large predators. The lion is the King of Beasts, once venerated as an animal god; the tiger is extolled for its beauty and strength; the wolf is a symbol of the wilderness. Yet at the same time these predators have been and still are persecuted to such an extent that they have vanished from much of their former range.

SERENGETI: A KINGDOM OF PREDATORS
GEORGE SCHALLER

MASAI MARA NATIONAL RESERVE in Kenya, where lions stand golden in the dawn, their breath raw and frosty. Mara, the Masai's 'spotted land', kingdom of all the predators. It was here during my first visit to East Africa in 1975 that I had watched a pair of black-maned lions striding slowly across the windswept plain, their sharply focused eyes turned skyward. Overhead, vultures wheeled effortlessly on outflung wings, pinpointing the position where other lions crouched flank to flank around a freshly killed zebra. The sight of those lions was to change my life.

Towards the end of 1976 I realized my dream of returning to the Mara. Everything looked different; I hardly recognized the rolling green plains that had so impressed me two years earlier. The Mara was as parched and thirsty as a dying man, the land brown and grassless, ugly in its starkness. Animals stood as if dazed, bony-hipped and listless, their coats rough and staring. For month after month the predators had an easy time of it, feasting on the wealth of carcasses that were everywhere made visible by the spiralling to earth of vultures. The buffaloes were particularly hard hit. Weakened by drought and disease, they were no match for the powerful prides of lions. There were times when even a single lion dared to challenge one of the old bulls, circling almost nonchalantly behind the heavy horns, waiting until the huge beast had exhausted himself with fruitless rushes before moving in to drag him down. The bellows of the dying animal would drift across the plains and before long the tawny heads of other pride members peered enquiringly from the bushes. So easy were the pickings that having suffocated the bull, the lions might gnaw away the soft flesh around the skull and then abandon the rest of the carcass.

But the Mara of the dry season is transformed during the long rains of March, April and May. Then the rolling plains are cloaked with a mantle of ripening red oat grass; well-worn tracks are obliterated among the chest-high sward. Great chunks of rock lie in wait, ready to

During the rainy season, when the vast migratory herds of wildebeest and zebras are absent from the Mara, lions survive on the resident prey species, particularly warthogs, buffaloes and topi. When food is scarce even the big males may be forced to try and kill for themselves.

punish a momentary lapse of concentration with a crunch so sickening that it can spew oil from a differential or suspend you half a metre above the ground. Marshy areas become quagmires into which heavily laden vehicles can sink up to their axles, and dry river-beds or luggas become raging torrents, denying one the chance of recrossing.

There is no doubting that the rains have a beauty all their own, the russet seed-heads swaying like a lion's mane in the breeze. But those grass-covered plains are never more treacherous than the day after a heavy downpour. As Martha Gellhorn noted in *Travels With Myself and Another*:

> This sinister stuff is called black cotton and combines the qualities of quicksand and chewing gum. Experience has taught everyone that the only way to handle it is to go around or go away.

I could have had no better ally in those early days than Joseph Rotich, a wise and keen-eyed Kipsigis driver based at Mara River Camp, where I was now living. Joseph was known in Swahili as *Bwana Chui* – Mr Leopard. But it was not only leopards that he found with consummate ease. Joseph knew the Mara like no one else. He taught me how to avoid the

A six-month-old cub of the Out of Africa pride tries to play with one of the pride males. Although pride males will sometimes tolerate this pestering, they are not as indulgent as the females, preferring to spend most of their time with the other males of their coalition and seeking out females only when they have a kill to share or when they are in oestrus.

more treacherous places, advising me when to turn back and where else I might cross during the height of the rains. More importantly, perhaps, Joseph showed me where to look for the lions that I so eagerly sought. Each pride has seasonal hunting grounds and favoured resting places, secret hide-outs where lionesses commonly give birth to their cubs. There were memorable evenings around the campfire when Joseph regaled me with stories about the Mara, an area he had known for more than twenty years. I sat enthralled as he talked late into the night, pausing only long enough to identify the sounds of mole crickets and fruit bats or to name individual lions and leopards that suddenly gave voice in the darkness. When Joseph spoke of lions, one place in particular kept recurring in his conversation: Musiara Marsh.

The marsh is a well-known landmark for all the drivers from the tented camps. It lies just inside the northern boundary of the Mara Reserve, on the east side of the Mara River. Animals are drawn to its green pastures throughout the year, and never more so than during the dry season from July to October, when the migration of more than one million wildebeest swarms in from Tanzania's vast Serengeti National Park. Each day, huge herds of wildebeest and zebra make their daily pilgrimage to the spring-fed waters that seep from the base of a low rocky ridge bordering the acacia woodlands to the north.

Having learned how to navigate the seasonal quagmires surrounding the marsh, I could now begin to concentrate on the animals, particularly the pride of lions within whose territory it fell. Joseph told me all that he knew about them, painting a vivid picture of the rich tapestry of life that defined the Musiara Marsh area. During the migration, the Marsh lions, as they were known, would lie in wait for the massed armies of wildebeest and zebra; tawny shapes concealed in the reeds. I watched spellbound at the sight of the five lionesses crouched together, ears flattened, muzzles raised, hind legs bunched ready to launch them from their place of ambush. Soon I could recognize many of the Marsh lions as individuals – essential for anyone hoping to learn more about their social way of life.

I never dreamed that, thirteen years after I took up residence in the Mara, I would still be living and working in the area or that I would be looking out across Musiara Marsh in the company of the Marsh pride. In fact sixteen of the lions lay crammed together in the shade of my vehicle. I could still recognize some of the lionesses; all were related. This is a feature of every pride: each one consists of a union of related females with successive coalitions of adult males, who are unrelated to the lionesses.

But the Mara is much more than its magnificent lions. It is timeless, yet forever changing. Each day I set off in my vehicle to explore a new patch of forest or rediscover a long-forgotten bend in the river, striving to capture with camera or pen a previously unrecorded moment. The passing years have given a fuller perspective on the Mara, highlighted by the constantly shifting relationships between the various species. While one population of animals flourishes, another declines, recovering some years later as drought or disease alters the degree of competition: nature's balance always achieving a new equilibrium. Certain years are remembered as being particularly good for finding cheetahs or leopards, punctuated by the times when wild dogs successfully establish a den. Observers are quick to focus on these localized fluctuations in time and space, and to pronounce upon them. Yet they are merely ripples in the wider scheme of things. Between the extremes of drought and flood there are the so-called 'normal years', years when the short rains and the long rains arrive and depart 'on time', fulfilling, perfectly, the predictions of sages, who bemoan the passing of an era when you could plan a safari around the onset or the end of the rains. In the midst of this ebb and flow of events, one particular year stands out in my mind. The drought-like conditions of 1983–84 were memorable not so much for the lack of rain as for helping to create exceptional leopard viewing. It was a great opportunity. I had waited six years for this moment.

Leopards have long been an obsession of mine, their solitary way of life so different from the social habits of lions. In the early days I had begged Joseph to help me in my quest to find and photograph leopards, hoping perhaps that there was some magical short cut to all those years of experience that the old safari guide relied on to reveal their hiding places. From Joseph I learned that the golden spotted cat is rarely the leopard that you see. Recognizing a leopard in all its many guises is what counts: a patch of unusual darkness in the grass, a spotted muzzle peering above the rock face as a vehicle passes by, a long dappled shadow disappearing down the trunk of a tree more than a kilometre away. That is the leopard you must try to find.

OPPOSITE: *Many of the lions in the Mara have been born and raised among safari vehicles and are quite accustomed to the sight and sound of people in cars. On the open plains, the Marsh pride sometimes rested in the shade of my vehicle.*

OVERLEAF: *A typical Masai homestead or* engang *consists of a tightly packed circle of thorn-bush enclosing a number of loaf-shaped houses with dung-covered walls and roofs. The cattle are enclosed within the thorn-bush at night to protect them from predators – particularly hyaenas, lions and leopards – and from cattle rustlers.*

The loss of cover precipitated by the drought of 1983 made it difficult for leopards to move about during daytime without being seen. Now was my chance to enquire more deeply into the secret life of one particular female. She was known as Chui – Swahili for leopard. Chui was the most tolerant leopard I had ever encountered, though our meetings were always brief. Her mother was a huge animal, shy and wary of vehicles, a survivor of the years prior to 1977, when Kenya finally closed all trophy hunting and banned the trade in wildlife products. The poaching of leopards for their beautiful coats had become endemic throughout Africa, with 60,000 leopards a year being sacrificed to the greed of the fur trade. Today, far from being an endangered species, the leopard is again thriving in some of Africa's wild places, and is seen by most people on safari to Kenya.

The scarcest of all the predators in the Mara is not the elusive leopard or the vulnerable cheetah. Africa's most endangered large carnivore is the wild dog, a species whose sociable ways are even more highly developed than those of lions. Like many visitors to Africa, I knew little about these strange bat-eared creatures when I arrived in the Mara. Shot and poisoned throughout their range, wild dogs have been penalized by man for the manner in which they disembowel their prey, being typecast as cruel and wanton killers. Today, epidemic diseases, loss of habitat and the depletion of their natural prey have compounded their plight, making them a rare sight. Time is running out for the wild dogs.

Ironically, many of Africa's parks and reserves were initially set aside simply because man had little use for them. But the Mara and its surrounds are different: this is a land of enormous potential. Wheat fields creep ever closer from the east, compressing the people and the wild animals on to the remaining land.

Traditionally the Masai landowners have always been tolerant of wild animals, relying on their cattle to nourish them, and a nomadic lifestyle to sustain their traditional culture. But Kenya's population will double in the next seventeen years, and the number of pastoralists roaming the lush grazing surrounding the reserve is growing by the day. As the population becomes more sedentary, there are signs of overgrazing and erosion, and predators wandering outside the reserve increasingly pose a menace to man and his livestock.Today the herdsmen are forced to compete with huge numbers of migratory wildebeest and zebra which crowd their pastures during the dry season. Understandably, the Masai resent this intrusion. In 1984, the government acknowledged their demands for the return of seasonal watering holes and grazing sites, degazetting 162 square kilometres of the Mara from protected status, reducing the size of the reserve by ten per cent. There is even talk of eventually erecting fences to separate man and his domestic animals from the wild creatures.

During the last fifteen years tented camps and lodges have proliferated in and around the reserve. Tourist vehicles at times encircle the lions, leopards and cheetahs that visitors have travelled from all over the world to see. Yet, despite everything, the Mara looks much as it did when man's ancestors first left the safety of the trees and wandered out into the savannah two million years ago. It is still one of the world's most spectacular wildlife habitats, with the majority of animal species represented in greater abundance than at any time in living memory. If I had only one day in Africa, I would spend it here, in the Kingdom of Lions.

OPPOSITE: *Lions often drink after a heavy meal, though like most other predators they can go without water for considerable periods as long as they avoid heat stress. A cat's tongue is not a particularly efficient ladle for drinking with and lions may lap continuously for a number of minutes.*

—1—

The Kichwa Tembo Pride

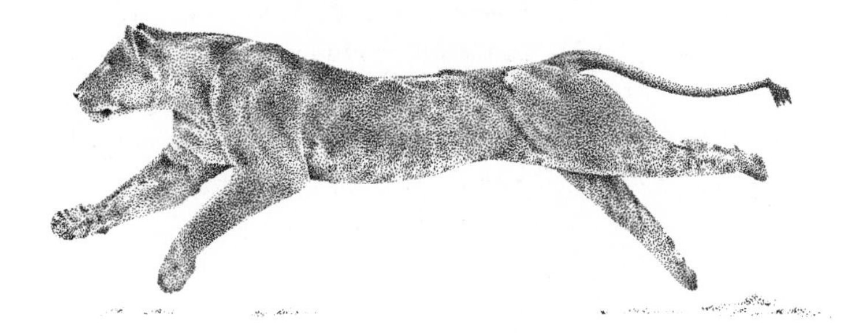

Lions are not animals alone: they are symbols and totems and legend; they have impressed themselves so deeply on the human mind, if not its blood, it is as though the psyche were emblazoned with their crest.

A Glimpse of Eden
Evelyn Ames

It had to be predators. The plains were alive with the tension that only they could impart. The blond grass-heads seemed almost to rustle with anticipation. A lone wildebeest galloped away, spurting forward as if startled, swinging its ox-like head from side to side, wary of the danger that might be hidden from view. The sound of the grass parting caused warthogs to stand rigid with alarm, noses thrust enquiringly into the air. Gazelles, impalas, Coke's hartebeest and topi all stopped feeding, signalling to every other living thing by their alert posture that a predator had been sighted.

The Kichwa Tembo lions paused, staring after the bull wildebeest with an intensity that was quite different from the way they usually cast their golden eyes across the game-filled plains. It was a look of recognition, an acknowledgement of the fact that wildebeest are favoured prey.

A few years ago it would have been unusual to see even one wildebeest on these plains, once the great herds from Serengeti and Loita had left the Mara in October. But in recent years Kenya's Loita Plains to the east of the reserve have fallen under the plough, transformed from open grasslands into thousands of hectares of wheat, as some of the Masai lease their land to farmers. Denied access to large tracts of their ancestral wet season pastures and calving grounds, the Loita's 90,000 wildebeest population has dwindled. Latest counts

OPPOSITE: *Topi are extremely vigilant and are often to be seen atop termite mounds, on the look-out for any danger. They are frequently the first to detect a predator on the move. Using the extra height of the termite mound is also a good way for a male topi to make himself visible to other males – an aspect of behaviour known as sight marking.*

Despite their size, male lions are capable of a surprising turn of speed, particularly when they sense the chance of scavenging from other predators' kills. They are perfectly capable of killing for themselves – and have to do so during their years as nomads – but once they have established themselves within a pride, they will take advantage of the lionesses' kills whenever they can.

reveal that only 20–30,000 wildebeest have taken to staying in the Mara all year round. For the last three years the old bull had chosen to remain on the Kichwa Tembo Plains, avoiding the barbed wire fences further to the east.

As the wildebeest galloped onwards, a small herd of topi ran towards the lions, marking the position of the predators. Tourists on an early morning game drive were stunned by this seemingly reckless behaviour. Surely it was suicidal to approach so close? But in fact the topi ventured no further than was safe, keeping just beyond the flight distance that each prey species knows will allow it to outrun a predator. In this way the topi indicated to the lions that they had been seen, and that it was futile for them to try and hunt here – but they kept the pride within view until it was safe to continue grazing.

Despite their interest in the wildebeest, the Kichwa Tembo lions had no need for food. All of them were visibly sated, for they had only just abandoned the carcass of an old bull buffalo, killed three days earlier. Since then, between them the ten lions had consumed nearly

400 kilogrammes of meat, leaving the remaining forty per cent of the carcass – bones, skull, horns, skin and rumen contents – to the vultures and hyaenas.

The attention of the five year-and-a-half old cubs was now firmly focused on a large male lion lolloping towards them. Blond Mane, as he was known, ran with the heavy swaggering gait so characteristic of his kind, mouth tense, ready to bully, to dominate. He had seen the antelope staring – watched as they scattered in alarm – knew from their behaviour that another predator was moving among them. Perhaps there was a fresh kill from which to scavenge a meal? As a full-grown male, Blond Mane exuded the confidence inspired by his enormous size and strength. He weighed over 200 kilogrammes, and measured three metres from his pink nose to the black tuft of his tail. No hyaena would dare to challenge him in this mood. The only animals Blond Mane feared were males from another pride, and it was unlikely that he would find them here. He was at the heart of his twenty-five square kilometre territory. If there was food, it was his for the taking. But as Blond Mane drew closer he found the lions to be the members of his own pride.

I watched the lions advancing. It was apparent that they were unable to recognize one another in the way that humans visually identify even a casual acquaintance. Sight alone does not provide a lion with sufficient information. Gestures and scent are all important. Stride, bearing, demeanour all help to confirm that a lion is friend, relative or foe. A lion is received according to the manner of its arrival and whether or not it is impregnated with the group's smell. This communal odour – passed from one lion to another by rubbing against each other – is the passport to pride membership, affording protection from unnecessary aggression.

The mothers of the five cubs veered towards Blond Mane, ears cocked, crowding him in that friendly yet assertive way which lionesses sometimes use to warn off adult males when cubs are present. The two lionesses greeted Blond Mane briefly, uttering an assortment of friendly *auu*'s and throaty rumbles, lashing their black-tipped tails from side to side. The cubs recognized the male. Each in turn – sometimes together – stalked towards Blond Mane, crouching as if in hiding, yet already seen, drawing him towards them, inviting him to play. Despite his adult size, Blond Mane was unable to disguise his youthfulness and responded exuberantly to the cubs' advances.

The cubs flung themselves against Blond Mane's tawny flank, colliding with the solid wall of muscle, sliding and tumbling off the thick curve of his back. Blond Mane broke into an ungainly gallop that sustained itself for just two or three massive bounds, egging them on. One of the four male cubs raced ahead of him, then flattened against the ground, waiting until Blond Mane was almost on top of him before rearing out of the long grass and embracing the huge maned head with outstretched paws.

I love to watch lions like this. So often, adult males appear aloof to the antics of young cubs or quickly put an end to any attempt to play with a growl or a grimace, baring long, ivory-coloured canines that, with one fierce bite, could crush a skull. Male lions greet and socialize mostly with the other pride males, seeming to prefer each other's company to that of the females and cubs. Their interest in the lionesses is largely prompted by the possibility of siring cubs or acquiring food from the females' kills. Yet the males are crucial to the pride's well-being, helping to define the extent of the territory and preventing intrusions from other

OVERLEAF: *During the long rains – from about the end of March to June – the plains are cloaked in a waist-high mantle of red oat grass. Because of the camouflage this provides, the Miti Mbili lionesses and their young cubs have stumbled upon two nomadic males. One of the lionesses has hurried away through the long grass, while another tries to keep the males from attacking and killing the cubs.*

males who might kill their cubs. These adult males – an alliance of two, sometimes of three individuals in this part of the Mara – are not related to the lionesses. Often they are brothers or cousins, though at times expediency forces them to forge links with unrelated nomadic males. To defend a territory successfully, pride males must act together.

The 'presence' of lions – the aura that surrounds them, so evident and alluring to the human observer – is not pure invention on our part. Nearly every facet of lion behaviour is tinged with confidence, a reflection of their size and strength, and their social nature. And it is the combination of size, strength and numbers that holds sway in deciding the balance of power among the predator hierarchy. Competition for food and space can be fierce, and is often translated into violence. Kilogramme for kilogramme, there is no question as to who is the king of predators: lions are so much bigger than hyaenas and wild dogs, leopards and cheetahs. Consequently, lions exhibit little of the nervousness often shown by cheetahs or wild dogs at a kill.

Living in groups helps lions and hyaenas protect their own kills or steal from others; both gain a significant proportion of their food by robbing the smaller predators. But competition is not limited to mere acts of piracy. Lions and hyaenas will also kill the young of other predators if they find them unprotected, although lions rarely devour their victims; they usually maul and then abandon them. Only a determined mob of hyaenas is capable of matching the singular strength of a lion; but even hyaenas know to keep a respectful distance when adult male lions are present at a kill.

Walking some distance behind the two lionesses and their cubs was a third lioness, distinguished by her age. She was square-jawed and broad-shouldered, her massive head creating an illusion of fierceness contradicted by the leanness of her body. The litheness that was so apparent in the stride of the younger members of the Kichwa Tembo pride was missing from the Old One's gait. Her coat was rough and staring, not sleek and shiny like those of her two adult daughters walking ahead of her. Her head lolled from side to side, spittle drooling from black lips. She could smell her daughters' familiar scent lingering on the tall stems of the red oat grass. Every so often the old lioness lay down, almost as if she were too tired to continue. She was, after all, well into her fifteenth year.

It was good to see the Old One again. I did not pause to check the identification file of photographs that I had accumulated over the years, documenting the unique pattern of whisker spots on each lion's muzzle. Torn ears, broken teeth, a part of a tail missing, the loss of an eye; all are useful for helping to recognize lions in the field, although none of these other physical characteristics is uncommon enough to be infallible. Yet despite the effects that time had wrought on the Old One's features, I knew this was her. She was the only survivor of the four adult lionesses of the Kichwa Tembo pride that I had learned to recognize in 1981.

In her younger days, the Old One was known to the drivers at Kichwa Tembo Camp for her exceptional hunting ability, and even as recently as 1989 she had proved herself still capable of killing a full-grown male topi. For a while she had continued to mate with the pride males, even though she no longer produced any cubs. But lately the Old One had stopped mating altogether. Her long and successful life was drawing to a close. Each day I searched for her, never sure if I would find her alive or dead. By now the drivers had christened her *cucu*, meaning grandmother in Kikuyu.

It really did not matter how quickly the Old One crossed the plain. Even if she lost her relatives' scent trail she would know where to look when she wanted to find the rest of the pride. She was familiar with every part of the Kichwa Tembo territory; it was the place of her birth. There was little need for her to hurry. Due to the quality of their protein-rich diet, lions can afford to move at a more leisurely pace than the wildebeest or impalas, spending much of the day and night resting, while the antelopes and gazelles must continue to feed.

The balance of power among predators depends on size and numbers, with male lions at the top of the hierarchy. Here, a lioness accompanied by an eighteen-month-old cub had killed a warthog, but the sounds of the dying pig soon attracted members of a clan of hyaenas. For a while, the lions and hyaenas fed at opposite ends of the carcass.

As more hyaenas appeared, the lioness sensed the inevitable and moved away. But the young male was desperately hungry and refused to move until he was attacked by the hyaenas.

In the end, the lioness could only watch helplessly as the hyaenas devoured her kill.

Bringing up the rear of the procession was the other pride male, known as Black Beard. He was the kind of lion that visitors dreamed of seeing. The drivers were forever asking each other in Swahili about the whereabouts of 'the *vichwa kubwa*', meaning 'the ones with the big heads'. They did not use English when exchanging information about the location of predators for fear of disappointing their clients. Visitors were always anxious to be shown lions and loved to watch the lionesses playing with their cubs; but above all it was the sudden emergence of one of these spectacular males that drew gasps of admiration from the onlookers. And nowhere among the kingdom of lions was there a finer lion than Black Beard. The sheer size of the animal; his magnificent mane; the length of those dagger-like canines – truly, this was an animal to be admired.

Now it was possible to distinguish Black Beard from Blond Mane, his four-year-old son. Black Beard's canines were the colour of old ivory, stained nicotine-brown with the passing years, while Blond Mane's were still clean and white. And Black Beard's nose was black as coal, not a youthful fleshy pink like his son's.

Black Beard drew alongside one of the safari vehicles and rubbed his bearded face against the rear bumper, revealing a thick rug of coal-black hair that hung like a curtain across his chest. Some of the visitors looked apprehensive and ducked their heads or quickly wound up the windows, while others tried to stand even further out of the roof-hatch for a closer look. Black Beard seemed unperturbed, ignoring the excitement his appearance had created. He had seen it all before, and anyway, the big lion did not associate vehicles with man – the creature that strode upright across his territory, the sight of which would have immediately caused him to run for cover.

He turned and arched his tail high into the air. A stream of urine, mixed with pungent secretions from his anal glands, spurted upwards, sparkling like fragments of glass in the sunlight, and trickled down the door of the Toyota Landcruiser. Ever since he was a tiny cub, vehicles had been a consistent part of Black Beard's world, as familiar to him as the orange-leaved croton bushes and fallen trees that he and his son scent-marked as they patrolled their territory.

Black Beard continued on his way, pausing every so often to stare ahead of him, drawing himself up to his full height so as to peer over the tops of the shoulder-high red oat grass. He could not see the Old One lying curled up, fifty metres away, but he knew she was there and, just as importantly, he knew who she was. The Old One had been a member of the Kichwa Tembo pride when, some four years earlier, Black Beard and his brother had successfully challenged the resident males and chased them away. Eventually, after a period of considerable upheaval for the pride females, Black Beard had mated her, and over the years he had sired a number of her cubs, including Blond Mane. The fact that the young male was still living among his female relatives was highly unusual. It was only because Black Beard's brother had died the previous year that Blond Mane had been allowed to stay.

Without Blond Mane to help protect the Kichwa Tembo territory, Black Beard would undoubtedly have been forced out by other male coalitions competing for breeding rights in the area. So the Old One and her daughters grudgingly learned to tolerate the continued presence of their young male relative, allowing him to share their hard-earned kills with a new generation of cubs.

OPPOSITE: *Black Beard, the old male from the Kichwa Tembo pride. In the northern Mara there are usually two or three adult males accompanying each pride. Almost all male lions leave the pride in which they are born at the age of about two or three years, forced out by the increasing hostility of their adult relatives, and their natural wanderlust.*

Old Man and the Old One mating. A male lion tracks a female in oestrus by her scent and will guard her, even from members of his own coalition, to prevent her from mating with another male. Repeated copulation induces ovulation in female cats, and a lion and lioness may mate up to 200 times over a three-day period.

The Old One did not look round at the sound of the grass being pushed aside. She sensed who it was. Black Beard reached down with his massive head, rubbing his face and mane against the old lioness's upturned cheek. With ears cocked and chin thrust forward the Old One responded, rumbling with leonine pleasure and revelling in the reaffirmation of her social being. That simple act of recognition seemed to strip away the years. For a brief moment the Old One looked alive, the weariness gone. Once more her life had meaning as a member of the pride.

On arrival back at Kichwa Tembo Camp that evening I sought out three of the older drivers. Over the years many of the drivers and camp managers have helped to keep track of the animals whose lives I have been following. This has proved invaluable in piecing together events that might otherwise have remained unrecorded. During 1986–88 I spent part of each year in Serengeti National Park in Tanzania, but whenever I returned to Kichwa Tembo, I endeavoured to hear at first hand what had happened to the various prides of lions during my absence.

Boniface Leva, Zablon Mbugwa and Francis Kugo have been at Kichwa Tembo for more than ten years. I first met them in 1981 when I visited the area in search of Scar, Brando and Mkubwa, the three male lions who had dominated the territory of the Marsh pride on the east side of the river in the late '70s. In 1981, the Marsh males were forced to flee for their lives, choosing to cross the Mara River rather than stay in the Musiara area and brave the youthful challenge of a new alliance of nomads. It was during this period that I found Old Man again, another exile from the marsh, and decided to leave Mara River Camp and base myself at Kichwa Tembo.

Zablon explained to me that after a period of stability in the Kichwa Tembo area, the inevitable transitions had taken place. There are always groups of young nomadic males and former territory holders seeking the opportunity to oust resident males. Each year the passage of the wildebeest and zebras through the Mara's spotted grasslands heralds an influx of these nomads. With such a bountiful supply of food, they are assured of sustenance as they try to claim breeding rights for themselves. By 1982, Old Man, Scar and Mkubwa – who had formed an alliance after Brando vanished – had grown old and, despite their impressive size and magnificent manes, they disappeared in the face of formidable opposition from younger males. Suddenly the Kichwa Tembo pride territory had three new males patrolling its grasslands. The sounds of their roars crashed across the luggas and through the thickets. Once again there were new scents being sprayed high on to the croton bushes and impregnating the game paths around the Sabaringo lugga at the heart of the Kichwa Tembo territory.

During this period, a coalition of seven male lions moved into the area surrounding Musiara Marsh. Nobody could remember seeing anything quite like this before: a vigorous alliance of brothers and cousins – such power, such manes. These four-year-old males were tough and mean, and the drivers from Governor's Camp soon nicknamed them Amin's Dictators, for they could chase and intimidate whoever they pleased. Their roars reverberated across the Mara River and along Rhino Ridge, instilling fear in the hearts of the neighbouring pride males. Invariably, stronger coalitions try to extend their influence so as to maximize their breeding opportunities. Sometimes they may gain control of two or even three different groups of females. Together, the seven males were able to dominate the territories of Musiara Marsh and Bila Shaka, which until the early 1980s had been united as the domain of the Marsh lions.

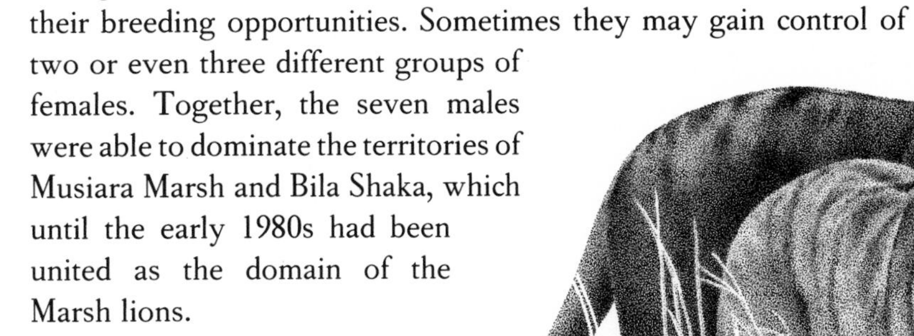

Nothing could stand in the way of these males. I well remember the day I found all seven of them feeding flank to flank around the bloated carcass of a bull elephant that had died deep in the heart of the marsh.

Eventually the coalition split up, four males remaining with the Marsh pride and three taking up permanent residence with the Bila Shaka pride to the east. Intense competition over food and females undoubtedly brought about the rift between the seven males and, a year or so later, two of them fell foul of the Masai's long-bladed spears as they attempted to ambush cattle near Campi ya Chui, just north of the Musiara gate at the entrance to the reserve.

Fortunately for the Kichwa Tembo pride, the Musiara males never encroached to the west of the river during the height of their reign. For a while the Kichwa Tembo lions prospered. Then one day the male known as Scruffy disappeared without trace, leaving Half-Tail and Rupture on their own. Within six months Rupture had also vanished. Throughout the next year Half-Tail struggled to hold on to his territory. But eventually, weakened by age and years of brawling, he was forced to abandon his pride or risk being killed by two males from the Serena area. If he was lucky, the old male might manage to form a new alliance with one or more nomadic males, and live long enough to win a new territory. But at his advanced age, his future looked bleak.

The loss of formidable male protectors is always a traumatic time for lionesses and their cubs. The disruption to the harmony of the Kichwa Tembo pride, caused by the arrival of the two new males – Black Beard and his brother – prompted three of the Old One's relatives to establish themselves as a separate group in the vicinity of the marshy land to the south of Little Governor's Camp, towards the end of 1986. They subsequently became known as the Out of Africa pride. The fact that the three lionesses were accompanied by twelve cubs, aged between two and eight months, undoubtedly hastened their departure.

Young lions become semi-independent at one and a half to two years of age; until this time their mothers do not come into season. But new males cannot afford to wait that long. In the interim they might well be ousted by other males, without having produced their own offspring. By attempting to kill Half-Tail's cubs Black Beard and his brother were helping to ensure that the lionesses quickly came into oestrus again and mated with them.

Having established themselves with the Old One and her daughters, the males continued to harass the three Out of Africa lionesses, determined to control both pride territories. Eventually, when their cubs were old enough to look after themselves, the Out of Africa lionesses accepted the brothers' advances.

By January 1990 six young nomadic males from the area surrounding Serena Swamp had become a potent force in the region. Throughout the period of the wildebeest migration, they had wandered along the edges of the Mara River, intimidating the resident males by their presence. Now they were unchallenged.

One morning the drivers from Kichwa Tembo Camp found the Serena males sprawled around a buffalo carcass. Thick bands of muscle outlined their shoulders and forearms, power boldly profiled beneath their tawny coats. Their breathing was slow and easy, their mouths held slightly ajar, revealing clean white teeth. A heavy breeze wafted away the flies that buzzed incessantly around their blood-stained muzzles.

OPPOSITE: *There are more than 350,000 Masai in Kenya and Tanzania, and many of them maintain their traditional way of life, roaming the plains with their cattle in search of fresh grazing. The Masai have always co-existed peacefully with the wild animals, unless lions or other predators attack their livestock. Some of Africa's finest wildlife areas are to be found in Masailand.*

Most young males spend a year or two as nomads after they are driven from the pride in which they were born. This is a difficult period, especially if a lion has no brothers and needs to form an alliance with other, chance-met nomads. Eventually they will oust the resident males from another pride and mate with lionesses to whom they are not related. This system helps to prevent inbreeding.

Not far away lay Black Beard's brother. The old male looked thin and weak with hunger. His proud face was mutilated, the left eye barely visible beneath his swollen cheek. He just lay there, his breathing laboured, not daring to move. There had been no hope of escape for the old male. All he could do when the young lions challenged him was to roll submissively on to his back like a young cub, peeing in terror, signalling that he did not wish to fight. But the young males were not to be appeased so easily. First they would teach Black Beard's brother a lesson – let him know that he was unwelcome in 'their' area. The brutality of their attack was shocking in its execution, blood-curdling in its intensity. They bit deeply into his hind-quarters, piercing his spine, raking their claws across his face and ripping mouthfuls of fur from his luxurious mane. When at last the six males turned and walked back to their kill, they pressed alongside each other, invigorated by their collective power.

The old male was finished. Where could he go? To the north lay the land of the Masai; beyond that was poachers' country. Across the river were the two Marsh males. Younger and stronger than himself, they would punish him if he dared to trespass, just as surely as these lions had done. As the sun dipped beneath the Isuria Escarpment, one of the young males walked back to where the old lion lay, with his massive head on his forepaws, as if in sleep. The young male bristled with renewed aggression, his mouth tense, ready to resume the attack. But it was too late. The old lion was dead.

Suddenly, a voice was missing from the chorus of roars that each night and morning marked the presence of the Kichwa Tembo lions, proclaiming their right to the pride territory. The long drawn-out rumbles, the bass tones of Black Beard's brother were absent. No further invitation to trespass was necessary.

For a number of months nothing untoward happened. The six Serena males had eventually split into two separate coalitions, one group of three taking over the Serena pride, the others claiming the Out of Africa pride. But surprisingly it was not these lions that decided to exploit the weakness that now pervaded the Kichwa Tembo Plains. That was left to the Kichwa Tembo pride's old rivals – the two black-maned males from the Musiara side of the river.

It was barely light when the Marsh males moved down to the edge of the river. The waters were icy cold, but nothing could deter the natural wanderlust of the two lions. They stood rubbing their great bearded faces against one another, staring across the river. The males were twins, distinguished only by the fact that one of them bore an ugly scar around his waist, where a poacher's snare had pulled so tight it had nearly cut him in two as he struggled to break free. Eventually he had managed to chew through the thick strands of wire, and a veterinary surgeon had immobilized him with a drug-filled dart and removed the snare.

There was no doubting the intentions of the Marsh males. They were always ready to try to expand their domain. The pride territory occupied by Black Beard was there for the taking. Having swum across the river, the two males shook the water from their manes and disappeared into the long grass. They prowled around the Kichwa Tembo area for the next three or four days, marking the territory with their own scent. Morning and evening their bellicose roars rolled across the plains, forcing the Kichwa Tembo pride to keep their distance.

In an effort to escape the unwanted attentions of these alien males, the Old One and her daughters fled with their cubs to the higher ground along the base of the Isuria Escarpment. The five cubs were jumpy and nervous. They sensed danger in the older lions' behaviour and stopped often to look over their shoulders at the slightest sound. But the Kichwa Tembo pride could not run for ever. Eventually the Marsh males caught up with them, a few kilometres beyond the Oloololo entrance gate. This time there was no chance to escape. The mothers of the cubs shepherded them into a nearby lugga, while the Old One stood to one side, watching the Marsh males approach, their manes blowing in the wind.

All male lions try to extend their influence beyond the boundaries of their own territory. The larger and more powerful a coalition, the more chance the males have of holding a territory and siring cubs. However, coalitions numbering five or more adult males usually split up eventually and control separate territories.

Blond Mane, the young Kichwa Tembo male, stared out over the long grass and saw the two Marsh lions moving purposefully towards him. By now he had grown accustomed to his role as protector of the pride territory. It was the only place he had ever known and without further deliberation he trotted out to meet the intruders. The drivers laughed at the brazen young male. They applauded his boldness, admired his audacity, yet knew him to be hopelessly ill-equipped to face the determined advance of the two more experienced lions.

As the Marsh males broke into a gallop, Blond Mane's confidence deserted him. These were no scrawny nomads. Within the space of one enormous stride Blond Mane turned and fled for his life. As he did so, he saw his father, Black Beard, running for the safety of the escarpment, a few hundred metres ahead of him.

Seeing their pride males fleeing, the Old One and her daughters hurried towards the place where the five terrified cubs cowered among the tangled tree-roots and thorn-bushes of the lugga. Suddenly the lionesses stopped and turned to face the onrushing males. It was almost as if they had shared one thought, and reached a common decision. They would not run from these strange males; to do so would mean certain death for their cubs. The lionesses' social nature, the bonding of kinship, years spent in each other's company had all helped prepare them for this moment. They had fought side by side many times in the past, confronting mobs of hyaenas and bands of young nomadic males. Now, together, they would fight again. They would contest the challenge posed by the two intruders, a subconscious recognition of the power of numbers in deciding ascendancy. Pitted against the raw strength and aggression of the Marsh males, the speed and agility of the three lionesses were not enough alone; but with the help of Black Beard and Blond Mane, it would be five against two.

The two younger lionesses charged as one, quickly gaining ground on the larger, slower Marsh males, who were stunned by the fury of their counter-attack. The Old One was there too, joining her daughters in battle, something she had done many times in the past. She showed a surprising turn of speed for her years, veering off to one side to harass the slower of the two males. The interlopers did not try to meet the challenge of the lionesses and ran for their lives.

Explosive grunts and coughs, issued as threats and appeasement, rent the air. Four times the trailing male was battered to the ground, twisting and turning, hind legs bunched up in a vain attempt to protect his rear end. He flipped on to his back to make best use of all four paws, lashing out with ivory-coloured claws that could easily slice through skin and flesh. But it was not enough to deter the lionesses. His brother, the snared male, made no attempt to help his stricken relative. He stood for a moment, watching as Black Beard and Blond Mane weaved their way back down through the rock-strewn escarpment, anxious now to join in the fray. But before they could do so, the battered male broke free and dragged himself away, trailing his left hind leg, which had been badly savaged by the lionesses.

By the time Black Beard and Blond Mane arrived on the scene, the Marsh males were distant specks on the horizon, hastening to reach the Mara River. Now it was the Kichwa Tembo males' turn to roar. The Old One and her daughters quickly added their voices, creating a barrage of sound, a mighty vindication of the power of the pride.

Later that morning visitors reported seeing two exhausted and wild-eyed males near the Oloololo gate, hurrying along the murram road. By evening they were once again deep in their own pride territory.

Not long after this, the injured male disappeared. His wounds had festered and were slow to heal, leading the drivers to believe that he must have died from the beating inflicted by the Kichwa Tembo pride. His disappearance left his brother, the snared male, with the unenviable task of defending the Marsh pride territory on his own.

As the long rains intensified during April 1990, the Kichwa Tembo pride were again on the run. Early one morning Black Beard and Blond Mane were challenged by the three Out of Africa males, not far from Little Governor's Camp. In the ensuing fight, Blond Mane suffered a deep wound to the side of his jaw. Even though the Out of Africa males eventually retreated to their own pride territory, the Kichwa Tembo lions left the plains and headed north over the Isuria Escarpment, where they were seen loitering within sight of one of the farms. For two weeks none of the drivers could find the pride, except for the Old One, who stayed close to cover along the Sabaringo lugga. By now she was too slow to run very far and was forced to hide, a fugitive in the land of her birth. Visitors sometimes heard her lonely voice piercing the night, roaring, roaring. 'Lions,' they would exclaim. 'We heard the lions last night. The Kichwa Tembo pride has returned.' But the drivers knew better. It was only *cucu*, the grandmother of the pride, searching for her relatives, calling for them to return from the escarpment.

—2—
The Isuria Escarpment

Happy the man, and happy he alone,
He who can call today his own:
He who, secure within, can say,
Tomorrow do thy worst, for I have liv'd today.

TRANSLATION OF HORACE III
JOHN DRYDEN

OVER THE YEARS, climbing the rock-strewn face of the Isuria Escarpment has become a personal ritual, a way of countering back-numbing months spent cloistered in my Toyota Landcruiser. I often long for the chance to stretch my legs, to feel the breeze on my face, to sharpen senses atrophied by confinement in a vehicle. Only on foot is it possible to capture the true intensity of wild places. To step around a rocky crag and catch a klipspringer unawares brings the antelope truly alive. To see three old buffaloes rise out of the long grass a hundred paces away and wheel to face you, noses turned into the wind, creates a spine-tingling awareness of their size and power. Flattened patches of grass, dried yellow by the weight and warmth of their enormous bodies, reveal the buffaloes' afternoon resting places. The sheer size of the grassy depressions distinguishes them instantly from the well-used look-out points created by the mountain reedbuck and defassa waterbuck, which also frequent the area.

The escarpment runs like a blue knife-edge along the north-western boundary of the reserve, transforming the scenery of the northern Mara. Home of majestic black eagles soaring overhead in their daily search for hyrax, and a refuge for the solitary leopard, it creates a pleasing sense of scale and perspective, enveloping the wooded grasslands within its crooked arm. Up here, 300 metres above the plains, one realizes what a wealth of bird and animal life lies concealed in the vast expanse of ripening red oat grass.

OPPOSITE: *The ground hornbill* (Bucorvus leadbeateri) *is the size of a turkey and has a deep, booming, almost lion-like call. Like the secretary bird, the ground hornbill is largely terrestrial and it spends much of the day scouring the grasslands for insects, reptiles, frogs and small mammals. The bird shown here is a male; females have a blue patch in the centre of the red wattle.*

African buffaloes are revered by big game hunters as one of the most fearsome adversaries. Living in herds which may number 500 or more, they are fast and immensely powerful, able to use their horns to deadly effect and capable of trampling a fallen victim underfoot.

Far below, the unmistakable silhouettes of three giraffes glide like tall ships across the plains. My binoculars mimic the vision of the tawny eagle that passes low overhead, revealing topi poised on earthern termite mounds. The mounds are grazed short by the topi and Coke's hartebeest, who seek out the nutritious star grasses colonizing the mineral-rich soil excavated by millions of termites. At the forest edges, delicate brown and white gazelles gather to feed on the close-cropped lawns created by the hippos' nightly forays on to dry land.

The tawny eagle is motionless in the eye of the wind, inspecting me with its piercing gaze before sweeping onwards. Familiar sounds mingle, elements that further define the Mara's unique character: the rushing of water cascading down the rock-strewn escarpment and the resonant bellowings of hippos wallowing in the river blend with the piercing whistles of Masai herdboys and the tinkling of cowbells.

I pause to examine the withered remains of a male impala. Fifty metres away a pile of crushed bone fragments and clusters of hair plucked from the antelope's coat have all the markings of a leopard's kill. The leopard must have surprised the impala during the night. I glance around for a suitable tree into which the leopard might have stashed the carcass. Maybe there was no need to place it out of reach, as there are fewer lions and hyaenas to compete for food here, outside the reserve. Perhaps it was simply too big?

ABOVE: *Leopards are masters of concealment, crouching motionless among rocky outcrops or croton thickets.*

OVERLEAF: *Many visitors prefer the forty-five minute flight from Nairobi to the five hour journey by road.*

Imprinted beside a pool of water are the pugmarks of a female leopard and her six-month-old cub. The escarpment is ideal leopard habitat, offering plenty of suitable caves for a leopard to seek shelter and rear young cubs in safety. But the female must remain constantly alert to danger, keeping a wary eye out for the Kichwa Tembo pride. Had she perhaps seen them as they clambered up the escarpment, and decided to abandon her kill?

This female leopard is shy of vehicles. Whenever she hears a car approaching, she crouches down behind the rocks, her young cub pressed close to her side. She has never learned to tolerate vehicles in the manner of the Kichwa Tembo lions; she always seems to feel exposed in their presence. As the daily procession of safari trucks bounces and rattles up the steep track, she invariably seeks the dense cover of the forest below the road. Occasionally a sharp-eyed driver might spot her as she snakes down the escarpment, her cub bounding alongside her. 'Leopard,' the driver exclaims with barely concealed excitement, and his visitors leap to their feet to glimpse the creature that they have longed to see.

The female lives a double life. There are times when she prowls around Kichwa Tembo Camp, slipping beneath the strands of electrified wire as darkness falls. Visitors returning from their evening game drive sometimes see her hurrying across the track leading into camp. And early one morning she was spotted departing with the cook's favourite cat clasped between her jaws.

ABOVE: *Troops of up to a hundred olive baboons roam the wooded grasslands in search of food. Baby baboons cling to the long hair on their mother's belly as she scours the plains and thickets.*

RIGHT: *Unlike wildebeest and zebras, baboons seem to be aware of the danger posed by crocodiles, recognizing them both on land and in the water. Baboons frequently cross the Mara River, leaping from rock to rock and even swimming when necessary.*

At times I lie in bed and listen to her calling, a repetitive, rasping cough, not unlike the sound of someone sawing wood. She is answered by the sudden, explosive bark of a bushbuck, its newborn calf huddled close to the ground. Baboons shift uneasily in the treetops, shrieking and barking, their alarm calls echoing around the camp, reminding me how uneasily the primate sleeps in the dark African night. Then quietness settles once again, and I know that the leopard has continued on her way, a shadowy figure melting away into the rocky hillside before first light.

I climb higher. It is almost four o'clock and the distinctive sound of Air Kenya's afternoon service from Nairobi drones on the wind. From out of the sky the veteran silver Dakota DC3 reveals itself. Within its steel canopy, visitors peer through glass portholes, savouring the excitement of being only minutes away from their first game drive in Kenya's premier wildlife area. The plane sweeps in low for landing, then rises sharply. With twin engines whirring the pilots curse the Thomson's gazelles and topi loitering perilously close to the murram airstrip. One of the waiting safari vehicles races alongside the strip, as the thirty-seater aircraft banks sharply to the south, then turns in a wide circle to try again. The driver leaps from the vehicle and hurls a rock in the direction of two male topi, who stubbornly hold their ground at the edge of the plain, so preoccupied are they with the priorities of the rut. They break apart as another missile sails over their heads. Skipping and prancing, bucking and tossing their heads with pent-up energy and aggression, they move away and the plane lands safely.

Flooded car tracks crisscross the plains, profiled by the harsh sunlight. After months of intermittent rain, the main road meandering along the base of the escarpment is torn and fractured. Day after day the run-off forms countless tiny rivers, in places turning soil into quicksand for the supply trucks servicing the camps and lodges. A troop of olive baboons crowds alongside one of these temporary streams, selecting lush shoots at the water's edge. With finger and thumb, each baboon deftly plucks individual pieces of grass, peeling away the outer sheath to reveal the succulent white stalk and tender green shoots they so obviously relish.

A kilometre from the airstrip, a cluster of vehicles marks the spot where predators have been sighted. Has the Kichwa Tembo pride returned to the plains? Perhaps it is the Old One, struggling to survive now she is apart from her pride. It is impossible not to worry about her. I never know if I am going to see her again.

To the east, the great plodding shapes of elephants are almost lost in the wide expanse of Musiara Marsh. Only they dare trespass deep within its soft green heart during the rains. Beyond them, a huge herd of buffalo spread like a black stain across the rich pasture, their numbers swelled by the seasonal arrival of chocolate-brown calves. They stand as if dazed, nearly 300 of them, their dark bodies numbed by the heat, rhythmically chewing the cud. A throbbing, drum-like sound points to where a pair of ground hornbills stalk through the head-high grass, duetting to one another as they search for grasshoppers, nestlings, lizards and frogs. Anything edible will do. Even terrapins encased in armour-plated shells are vulnerable to the pick-axe bills of these giant, red-wattled birds. A third hornbill appears, scruffily attired, with dull yellow wattle denoting its immaturity. The youngster waddles towards it mother, who has just caught a huge long-horn grasshopper with purple wings. It solicits the prize from the female – distinguished by a patch of navy blue in the centre of her wattle.

The sound of an axe biting into dry wood rings out across the hillside, drawing my thoughts back to the escarpment. In the distance two men, cloaked in blood-red *shukas*, stand stork-like beside a solitary tree. Chips of bark and wood mixed with half-dry grass lie piled into a small heap at the base of the tree, belching acrid smoke. The two men are quite

ABOVE: *During the rains, only the elephants, buffaloes and waterbuck can safely venture into the lush heart of Musiara Marsh in search of grazing. The Mara has a large elephant population – approximately 1,500 in and around the reserve; in 1984 some 400 elephants moved here, seeking refuge from heavy poaching in the northern Serengeti.*

OPPOSITE: *A view from a balloon. Some people have voiced concern about the impact of hot air balloons on the wildlife of the Mara, particularly with regard to elephants. In fact, a balloon trip only lasts an hour and, as long as the balloons do not fly too low over the animals, it is the perfect way to view the Mara's spectacular scenery.*

different in appearance. One is tall and slender, with finely sculpted features. The other is short – almost squat – with a less open bearing. Yet both are unmistakably Masai. We greet, as the men back away from the tree, ignoring the angry bees buzzing around their faces. Slung across the naked shoulder of one of the men is a freshly hewn bow, the smooth young wood ivory white. It is honey they are after. My old friend Joseph Rotich always enquired as to the state of the honey hunting whenever he visited the Mara. He would purchase it by the litre, spooning out great dollops and chewing the wax to a pulp before spitting it out. Nothing tastes as fresh and sweet as wild honey.

Returning to the tree, the men hack a bowl-sized gash into the underside of one of the thicker branches, exposing the dark hollow at its centre. This will encourage the bees to

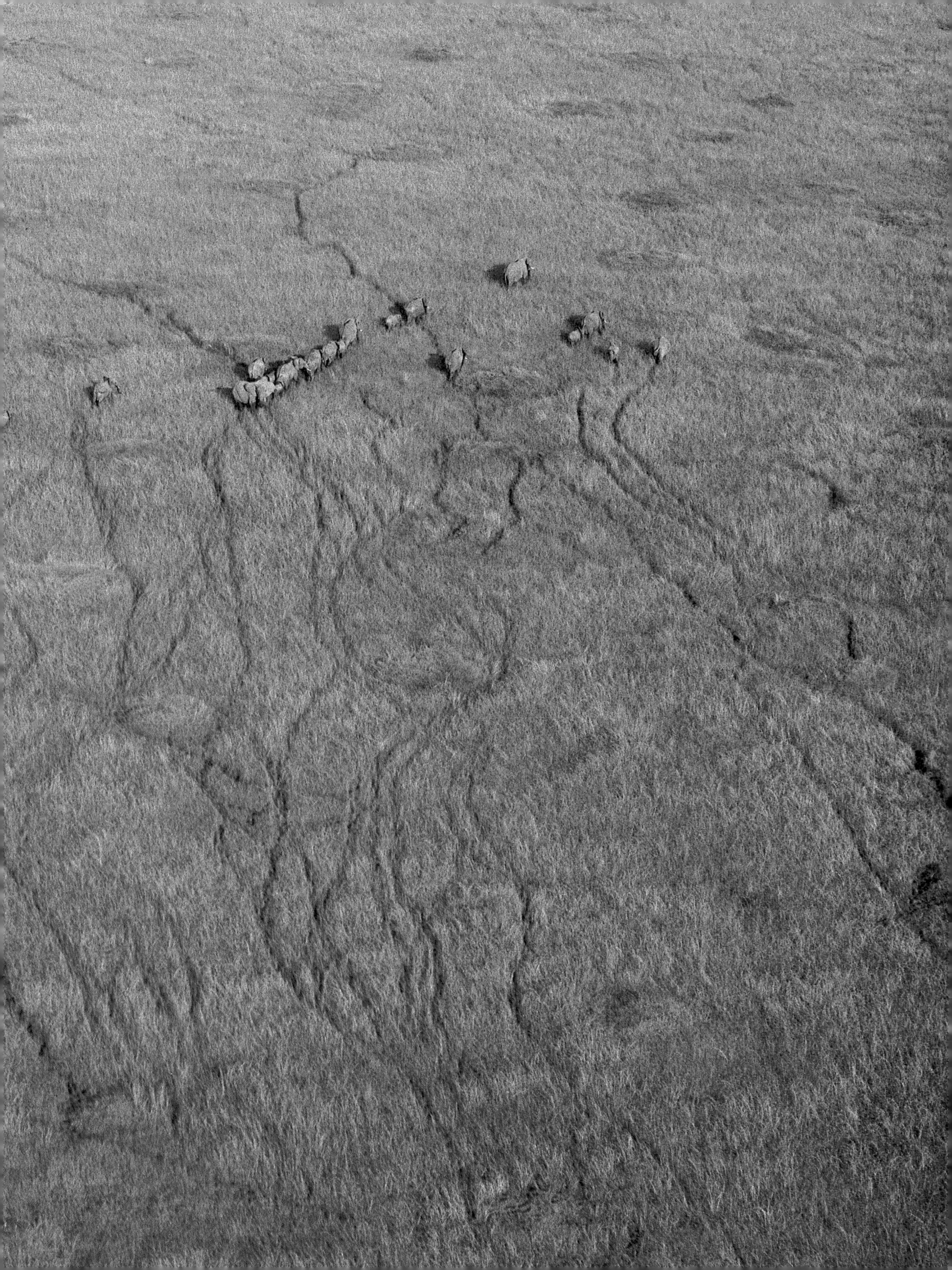

swarm here in the future. As I depart, the honey hunters head north, flighting an arrow towards a lone stump fifty metres on. I hear the *thwack* as its sharp metal point bites deeply into the wood. Bows and arrows are favoured by the Masai as protection against their sworn enemies, the WaKuria, who live astride the Kenya–Tanzania border. The WaKuria are tough people: cattle rustlers and sometime poachers, if one is to believe the rumours. Even the proud Masai fear them. They are said to be smokers of *bangi* (marijuana) and emboldened by the leaf are careless with life – their own as well as that of others. When raiding for cattle, they too come armed with bows and arrows, or carry AK 47 automatic rifles appropriated from the Uganda–Tanzania war.

Descending from the escarpment, I spot the dozen bull buffaloes that by day frequent the plain in front of Kichwa Tembo Camp. These massive wild cattle follow an unvarying routine. At night they move right up to the electric fence encircling the perimeter of Kichwa Tembo, seeking the cool, muddy water that seeps from pipes draining the camp, creating a perfect wallow. Before the fence was erected, buffaloes and hippos regularly wandered into camp during the dry season to graze alongside the tents at night. I watch the bulls plodding single file towards the water, gathering up others from out of the long grass. There is a spring to their heavy bovine gait, an energy fired by the expectation of the wallow. When they reach it the bulls grunt and belch, their heavy bossed horns crunching head to head as they push and heave among themselves. A party of warthogs, deferring to greater size and strength, grudgingly give way, hauling themselves out, shiny and dripping wet. One of the bulls immediately rolls on to his back. Then digging the point of his horn into the mud, he twists his massive neck, enlarging the hollow.

In twos and threes the buffaloes take their places in and around the wallow. Soon each familiar site is occupied and the bulls lie, eyes vacant, staring into the distance as they mulch and regrind the grass consumed earlier in the day. The economy of ruminating suddenly seems obvious. In a world inhabited by predators, feeding with head bowed to the ground is a dangerous necessity for the grass-eating animals. By re-chewing their food, ruminants are able to extract greater value from it in a place of their own choosing. Nevertheless I recount the bulls each day, knowing that the time will surely come when the hunger of the rainy season will force the Kichwa Tembo pride to seek them out. One bull has a broken horn and torn left ear. Another has no tail and worn teeth. I wonder: which one will it be?

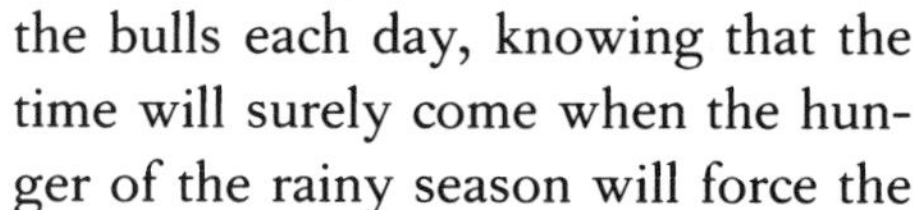

A newborn elephant calf weighs about 115 kilogrammes and stands just under a metre tall at the shoulder. It is strong enough to follow the herd within a few days. During its first year of life it will grow thirty centimetres or more – a first-year calf can be identified because it is still able to pass freely underneath its mother's body.

Later that evening I walk out on to the plain and peer into the darkness. On the ragged crest of the escarpment I can see a red necklace of fire spreading slowly eastwards from the place where I met the honey hunters. Had it been the dry season, the fire might have raged hot for day after day, destroying hundreds of hectares of forage.

Far off, the faint sounds of the Marsh lions greet my ears. For a moment the buffaloes fall silent, listening to the echoes of the predators. But the danger is far away. The moment soon passes, and the bulls resume their noisy wallowing.

From the north, rolling down the escarpment, the roars of the Marsh lions are answered. It is the Kichwa Tembo pride, adding their own deep voices to the night, telling others who they are and that they are nearby. Is there menace in their voices, I ask myself? Is it all in the hearing, a question of experiences past? What do they know? Do the Marsh lions fear the Kichwa Tembo pride when they hear them calling?

Sixteen lions lay at ease in the warmth of the dry murram road near Mara River Camp airstrip, appearing pale and ghostly in my car lights on this rainy April morning. The tattered ears and protruding lip acknowledged the presence of Short Ear, oldest of the Marsh lionesses. Some of the younger cubs looked gaunt and leggy as they walked ahead of the car, the tops of their pelvic bones protruding beneath their loose skin. More than a month had

ABOVE: *Lions often scratch at the base of a tree with their claws, leaving a visible tear in the bark. This is a form of territorial marking and also helps to keep the claws sharp.*

OPPOSITE: *Although a lioness stays within the pride territory to give birth, she selects a secret place, well protected from other predators and the rest of the pride.*

passed since I last saw the Marsh pride. Getting around during the height of the rains had proved so difficult that I had been forced to stay close to Kichwa Tembo, waiting for the weather to improve. Though the drivers told me whenever they saw the Marsh lions, there were times when the big cats simply melted into the long grass or remained invisible in the thorn-spotted landscape to the north of the marsh.

The territory occupied by the Marsh lions is one of the smallest in the Mara, providing a year-round supply of the species that the lions most frequently prey upon, such as warthogs, buffaloes, topi and zebras. It is these resident species – not the migratory wildebeest – that ultimately control the number and density of lions that can survive in a given area. But as with all pride territories, the range of the Marsh lions was not limited by arbitrary points on a map. Each territory is fluid in its extent, varying according to seasonal changes in prey availability, balanced by the fortunes of individual pride members and the density of lions in the surrounding areas. But the sudden loss of a key pride male or the impact of deluge or drought on prey abundance are just two of the factors that might bring about a change in the established order.

Later that same morning I found the lions huddled together in the shade of a dense patch of bush, not far from a Masai manyatta. It was hard to believe that there were sixteen lions in the long grass. A whole pride concealed in the space of a few square metres. But this was the only

Lionesses give birth to up to four cubs at a time, after a gestation period of 110 days. Here, one of the Marsh lionesses with her cubs leaps across the stream that gives rise to Musiara Marsh. Lions tend to abandon the wet areas during the rainy season, preferring the higher, better drained areas favoured by warthogs and topi.

way to survive among the Masai and their cattle. Keep a low profile and hunt only at night when the herdsmen were safely confined to their thorn-bush enclosures. The Marsh lions could no longer afford to lounge around in the open. To see them slink away from a group of Masai was to realize man's impact as top predator on the plains. A combination of instinct and learning had taught the lions to fear man like no other animal.

Sprawled with Short Ear and the other five adult lionesses were nine cubs of various sizes, all of them related to the old female. The smallest were her four great-grandcubs, born during the abundance of the migration. Lying in the long grass some thirty metres from the rest of the pride was the father of the four cubs, the old male with the snare wound. He was a magnificent, ginger-maned lion who three years earlier had taken over the Marsh pride territory with his brother. Now his brother was dead and the old lion appeared ill at ease, constantly raising his enormous head above the grass to stare about him; his usually confident mien had evaporated.

Short Ear had been mother to countless cubs, many of whom had perished before the age of three months. The same is true for the offspring of all lionesses. Predation by hyaenas and leopards, starvation and conflict at kills, abandonment and the inevitable episodes of infanticide by wandering coalitions of males – all these account for the fact that more than sixty per cent of lion cubs fail to reach adulthood. But during the ten years that Short Ear had been sexually active, she had succeeded in raising perhaps eight cubs to maturity. Of these, all the males had been forced to leave when they were two to three years of age, eventually challenging pride males from other territories for the right to mate with their lionesses. This helps prevent inbreeding within lion society. The majority of the female cubs had remained part of the Marsh pride – food and space permitting.

Like most lionesses, Short Ear had lived her life in the area where she was born, acquiring by association with older female relatives the knowledge of the best sites for giving birth and ambushing prey, finding water and shelter. But this is not always possible. Lionesses can live up to fifteen years, twenty even, in less competitive lion habitats. There are times when there is insufficient food to sustain yet another generation of female offspring within the pride territory. In these circumstances, some of the young females must leave, to live the life of nomads as they search for a territory of their own. But life for nomadic lionesses is exceedingly tough. A territory is life itself to a lion. Without a place of permanent residence and the protection of pride males, it is almost impossible for lionesses – particularly if they are on their own – to raise cubs successfully. Lionesses such as these continually become pregnant by roving pride males and nomads. But the offspring of these casual liaisons are invariably killed by other males.

Throughout the long rains the Marsh lions forsook the Musiara area. Slopping around in the water-logged soil surrounding the marsh was no place for them. It was as if they knew that no other lions would seek to establish themselves in the marsh during their absence, except perhaps for nomads, who could easily be driven out once the dry season began towards the end of June. From time to time the Masai would tell me that they had seen Short Ear with some of the cubs, and that the old lioness was even more recognizable now that two-thirds of her tail was missing. Meanwhile, a number of sub-adult males had been forced from the pride.

With so many cubs to feed, every kill made by the Marsh lionesses was keenly contested. Buffaloes are less plentiful to the north of the marsh and for a while the pride subsisted almost exclusively on warthogs. The mood among pride members was often explosive, and the scrawny cubs showed little desire to play. The lionesses had become increasingly hostile towards the old male. There were times when they drove him away as he lumbered in to claim a share of their hard-earned kills. I could not help feeling that it would not be long before he was forced to relinquish his tenuous hold on the Marsh pride territory.

—3—

Khali the Lioness

Man is a brief event on this continent; no place has ever felt older to me, less touched or affected by the human race . . . the wild animals belong here.

TRAVELS WITH MYSELF AND ANOTHER
MARTHA GELLHORN

IN THE PAST, at the beginning of each dry season, Short Ear and her relatives had quickly re-established their claim to the marsh, which provided the pride with the easiest hunting grounds in the whole of the Mara. From July to October the migration of wildebeest and zebras invaded the rich pastures surrounding the dense reed-beds, galloping down from the high plains and acacia woodlands to drink at the spring. But this year was to be different in the marsh.

As the rains faltered, the drivers from Governor's Camp once again felt free to scour every inch of the Musiara area, searching for Short Ear and the other members of the Marsh pride. They occasionally saw them loitering in the shade of a dense patch of forest just to the north of the spring. By now, the Marsh pride was a pale shadow of its former strength and numbers. The cohesiveness that had been such a feature of the pride seemed to be disintegrating. The old lion with the snare wound had disappeared and some of the lionesses had taken to wandering off on their own, deserting their cubs for days at a time as they searched for food. The younger cubs stayed close to Short Ear, feeling a sense of security in the company of the old lioness. Something was preventing the Marsh pride from reclaiming the heart of their dry season hunting grounds. The reason soon became apparent. And it had nothing to do with their rivals of old, the Kichwa Tembo pride, who were always ready to trespass around the spring whenever the Marsh lions were absent.

OPPOSITE: *Lions do sometimes climb trees, though they do not possess the grace and agility of leopards. Particularly during the rainy season, when the red oat grass grows almost as tall as a lion, they take advantage of the elevated view to search for prey or other pride members, to escape the attention of biting flies and even to avoid a herd of angry buffalo.*

When fighting, lions often rear up on their hind legs to enable them to lash out with their forepaws and bite at each other. These young Marsh lionesses are play-fighting, invigorated by the cool, muddy conditions of the flooded plain between the Miti Mbili lugga and Musiara Marsh.

Living to the east of the marsh, in the adjacent territory, were the lionesses of Bila Shaka. They were close relatives of the Marsh pride, but no longer friends. In 1981, three young lionesses – one of whom was Short Ear – were forced out of the Marsh pride, when a new alliance of males moved into the area. But they did not go far, establishing themselves in a narrow corridor of wooded grassland between the river and the marsh, leaving the Musiara area divided. Their older relatives – the original Marsh lionesses – were denied access to the marsh, and the drivers renamed them the Bila Shaka pride. These lionesses were the most localized of all the lion prides in the northern Mara and therefore very popular with the drivers from the camps and lodges. This was reflected in their name, which is Swahili for 'without doubt'.

By the time I picked up the tangled threads of their story, in 1989, there were four adult lionesses in the Bila Shaka pride. All were very different in appearance. One of the females had a coat the colour of rich toffee and was known as Brown. She had a fierce look about her, her dark furrowed brow and piercing eyes reminding me of Notch, who had lived along the Miti Mbili lugga many years earlier. In fact, the two lionesses shared some of the same genes. Yet for all her fierceness, Brown was not the most recognizable of the four lionesses. The most striking of them was noticeably older than the rest. Her pelage was grey and grizzled, and she had that gnarled, time-weary look that comes to every lioness with advancing years.

A lioness playing with her sub-adult relatives. At one and a half to two years of age, lions are big enough and strong enough to involve themselves in all aspects of pride life. They increasingly take the initiative in stalking and killing prey, and as a group are prepared to challenge an animal the size of a buffalo.

Grey Coat, as I called her, was a marvellous hunter, at the height of her skills. Years of experience had taught her precisely when to freeze in mid-stride and when to stalk forward just a few metres more, before launching her deadly charge. All of this natural wisdom Grey Coat possessed, combined with breathtaking speed and endurance.

In December 1989, Grey Coat, Brown and her sister had given birth to ten cubs within a few days of each other: five males and five females. Such synchronized births among lionesses within a pride are common, given the propensity of new males systematically to try and kill any young cubs sired by the previous coalition. Older cubs, particularly young males, are often exiled from the pride at this point. Without the dependency of cubs, the females quickly come into season again. But though they willingly mate with any new males wandering through their area, it is usually a number of months before they conceive. This period of grace helps ensure that the 'fittest' of the various male coalitions consolidates its claim to the area before the lionesses become pregnant.

The fourth lioness in the pride had given birth some months after her sisters and cousin. But instead of moving away and keeping her four cubs carefully hidden from view for the first six weeks (as is normally the case), the lioness gave birth in a dense croton thicket, close to where the other lionesses and their ten boisterous cubs often rested. The cubs' eyes had barely opened when the rest of the pride found them; the benefits of synchronized births soon

ABOVE: *The male in the picture has appropriated a warthog kill from the lionesses of his pride. At first, two of the females hurried away with their three-month-old cubs, but later the male allowed the cubs – who were either his own or his brother's offspring – to feed with him.*

OPPOSITE: *Lion cubs are kept hidden for the first six to eight weeks of their lives, but even so sixty per cent do not survive, whether because of infanticide by new pride males, starvation, desertion or predation. By the time they are the age of this cub (about eight weeks) they will have started to eat meat and to socialize with other members of their pride.*

became apparent. Despite her efforts to keep the older cubs from toying endlessly with her young, the lioness found it impossible to keep them at bay. Competition at kills was fierce and always favoured the larger cubs. Within a few weeks only one of the four cubs was still to be seen. He, too, disappeared during the long rains, when food was at its scarcest.

The core of the Bila Shaka pride territory is the Miti Mbili lugga which stretches northwards from Governor's Camp airstrip towards Leopard Gorge and Emarti ya Faru, the area frequented by the Gorge pride. Invariably, one finds the Bila Shaka lionesses and their cubs sprawled along its southern reaches, occupying a corridor of open ground surrounded on three sides by dense croton thickets and tree-lined water-courses. The lugga acts as a crossroads for the animals grazing on the surrounding plains. There is water to be found in scattered pools along the narrow drainage channel during the dry season. Here, wedged among the leafy shadows, the lions rest up during the hottest part of the day.

Male warthogs use their impressive tusks to defend themselves from attack, and to spar with other males. They are an important source of prey for lions in the Mara, especially when the vast herds of migratory animals are absent. All the larger predators prey on the piglets which are born during the short rains, when there is plenty of short green grass to feed the lactating females and their young.

Wildebeest and zebras, buffaloes and topi all at times make the dangerous trek to the pools. From the cover of the green and orange croton bushes, the lionesses might sneak forward, crouching in the tall grass at the edge of the rolling plains as the sun dips below the escarpment, or wait until darkness when their acute night vision makes it easier to ambush their prey.

By June 1990, the Bila Shaka lionesses were finding it increasingly difficult to kill sufficient prey for the whole pride. It was not just the ten cubs that the lionesses had to provide for. The two large pride males who had fathered the cubs had doggedly tracked the movements of the lionesses throughout the lean months of the rainy season. Visitors sometimes expressed contempt at the bullying tactics adopted by the two lions who, due to their far greater size, often managed to steal smaller kills, such as warthogs and buffalo calves, from right under the noses of the lionesses. But once they had eaten their fill, the males sometimes allowed their ten young relatives to feed from the remains; but the lionesses – who were not related to them – were usually excluded. They would have to hunt again to satisfy their own requirements. Unfair as it might seem, this division of labour between adult males (as defenders of the territory) and females works to the advantage of the pride.

As the dry season of 1990 continued, it became common knowledge that there was a solitary lioness prowling around the edges of the marsh. It was equally apparent from the way she assiduously avoided contact with the four lionesses that she did not belong to the Bila Shaka pride. Khali, as she was known (the word means 'angry' in Swahili), often frequented the vicinity of two weed-choked pools hidden deep within a patch of forest at the western edge of the reed-beds. She was a member of the Marsh pride and thought to be one of Short Ear's daughters. As an adult, she had remained with her female relatives within the territory of her birth. Each night, as the sky edged towards darkness, Khali would call to the rest of her companions. But Short Ear and the other members of the Marsh pride remained to the north, preferring to compete with the smaller Gorge pride for food and space, seeking shelter from the powerful Bila Shaka pride males among the grazing lands of the Masai.

By now Khali was heavily pregnant with her second litter of cubs, and seemed determined to give birth to them in the area she knew best. Nobody was sure who had sired the cubs. Had it been the final act of the old Marsh male before vacating the pride territory? Or was it one of the Bila Shaka males, who now patrolled the boundaries of both areas? For the moment there was no one to challenge them from roaming wherever they pleased. They would undoubtedly have mated with any 'unguarded' lionesses they met.

I often found the two males lying near to where Khali rested, or watched as they brazenly stole from her kills. But despite the intimidation and the solitude, Khali refused to retreat. The marsh was an ideal hunting ground now that the rains had ceased. Warthogs flourished here, wallowing in the drying pools between the riverine forest and the marsh. And the number of waterbuck had risen dramatically these past three years.

The increased rainfall had provided abundant food for the shaggy-coated waterbuck, who looked slightly out of place in the dry heat of Africa. Hunters and safari guides often maintain that lions disdain the flesh of this large antelope. Delicate human palates find the meat to be greasy, with an unpleasant smell. The assumption that lions think likewise seems egocentric at best. The most likely reason why waterbuck do not form a more sizeable part of the predators' diet is that they are wary and difficult to stalk, often retreating to water when threatened by pack-hunting animals such as wild dogs and hyaenas. The majestically horned males have the added advantage of being highly dangerous adversaries; one slip and a lioness could be fatally gored. Leopards, which have never been known for their fastidiousness, quite frequently take waterbuck calves. These youngsters spend the day 'lying out' in dense thickets among the riverine forests, being far too conspicuous and vulnerable to accompany their mothers to the open places where the herd is feeding. But leopards, their keen senses attuned to such cryptic behaviour, occasionally find these young calves as they prowl through the Teclea thickets.

Despite her nightly calling, the only lion sounds that Khali regularly heard were those of the Bila Shaka pride. Fortunately Grey Coat and her relatives seemed content to leave her be. A single lioness posed no threat to them, and there was little to be gained by trying to drive Khali away.

With the departure of most of the wildebeest and zebras from the plains surrounding Miti Mbili during September, the Bila Shaka lionesses were forced to wander further afield. The marsh now lay vacant. Only Khali remained from the old order. There was nothing to prevent the Bila Shaka lionesses from appropriating the area that, ten years earlier, had been an integral part of their territory. Grey Coat alone among the four lionesses was old enough to remember that period, and it was she who now began making increasingly frequent forays into her old haunts, lying in ambush around the forest that Khali had chosen as her retreat. Soon the old lioness was joined by the rest of the pride, forcing Khali to live the life of a fugitive. But still, as the birth of her cubs grew nearer, she refused to move away.

September merged imperceptibly with October, bringing a sense of expectancy as the time of the short rains approached. The rains were never predictable. Would they be early or late, sparse or heavy; might they blur with the long rains, obliterating memories of the previous year when the short rains failed? It was impossible to guess what would happen any more. The Masai simply turned their gaze to the sky and shook their heads with frustration. The plains were alive with the seasonal bloom of young warthogs and topi, providing a welcome source of easy food for Khali during the last days of her pregnancy. Surely the rains must be imminent? The waste-paper plants and fireball lilies did not wait for an answer. As soon as the first showers speckled the plains they erupted from the earth, covering the grasslands with clusters of white and the termite mounds with globes of red.

Each day, at around the time when visitors were heading out on their afternoon game drives, it poured with rain. The darkness gathered early and the morrow promised to be difficult for the mini-buses as they slipped along the muddy tracks without the benefit of four-wheel drive. But it is never wise to lie in bed, on the assumption that it will be dull and grey. The mornings are invariably washed with sunshine in the Mara, the plains green and radiant from the previous day's showers. Visitors return for breakfast ecstatic, having glimpsed a leopard just outside camp. Others count themselves equally fortunate to have watched two male cheetahs pull down a young topi on the glistening plains near Aitong Hill, where I often search for the elusive wild dogs.

On the fifth day the sky brightened and clouds vanished, swept aside by gale-force winds that we all knew would keep the wet weather at bay. For a while the rains faltered, interspersed with dry periods, followed by more rain, encouraging the migratory herds to tarry a while longer. I had never seen so many animals around the marsh at this time of the year, with wildebeest and zebras hugging its green fringes, seeking out the fresh water at the spring. A huge herd of buffaloes munched their way through the heart of the reed-beds, with elephants strung out like so many huge grey rocks as they marched solemnly to water.

With so much prey available, the Bila Shaka pride hardly moved from one day to the next. Occasionally I found them lounging on top of the termite mounds bordering the marsh, waiting for the sun to wane or for an unwary wildebeest to stumble towards them. But so sparse was the cover to shield them from their prey that they mostly hunted in the cool of the night.

OPPOSITE: *The short rains generally last from mid-October until December. The mornings are glorious with blue skies, but it usually rains in the middle of the afternoon. Blossoms spring up and the plains are covered with small white flowers known as waste paper plants; fireball lilies and pyjama lilies are also common.*

OVERLEAF: *Adult elephants are one of the few creatures completely safe from attack by lions. Living in large matriarchal family groups helps protect the calves until they are old enough to fend for themselves.*

Lionesses do most of the hunting; at 110–130 kilogrammes they are smaller and more mobile than the males and, lacking the ostentatious mane, can stalk their prey more easily. By living in groups and acting together, lionesses are able to defend their pride territory against other groups of females and have a better chance of protecting their cubs against male intruders.

Half a kilometre to the east of where the Bila Shaka pride rested stood three huge trees, like a signpost at the edge of a peninsula of forest which reached out into the middle of the marsh. The trees had been a useful landmark in the past, and I often found the Marsh lions resting beneath them with their cubs. Each tree was different in shape and character, and each over the years had provided the lions with shade or a safe retreat when danger threatened. Khali knew the area well, and I thought she might choose the vicinity of the trees as the birthplace for her cubs. But during November the wildebeest and zebras drifted south towards favoured river-crossing sites on Paradise Plain; it was time for the herds to return to the Serengeti. As prey became ever scarcer, the Bila Shaka lionesses increasingly harassed Khali, keeping her on the move.

When I discovered the place that Khali had finally selected, I was dismayed. But perhaps she had no choice. Exposed at the edge of the marsh, a solitary tree lay on its side, providing scant security for Khali's four tiny offspring. Vehicles were able to drive right up to the tree as tourists attempted to photograph the cubs. The only respite came when the visitors returned to camp for breakfast or dinner. But even then, there were always the picnickers from Mara Buffalo Camp, or local residents camping along the banks of the Mara River, who arrived during the middle of the day. Khali was remarkably tolerant of all the attention, patiently guarding her cubs. Then one afternoon, the buffaloes found her.

Buffaloes are non-migratory, though they move over a large home range. Like warthogs, they are an important prey species for lions in the absence of the migratory herds. Most calves are born during the long rains, and are preyed on by hyaenas as well as lions, although the powerful adults will usually try to defend one of their number against attack.

Khali could see the herd eating their way towards her. They moved together, shoulder to shoulder, a massive black feeding front pushing resolutely towards the forest edge, where Khali watched from beneath the fallen tree. To begin with, the buffaloes seemed unaware of the lioness. A pair of hammerkops danced and fluttered at the feet of the herd, darting in and out of the thicket of sturdy legs to snatch frogs and insects. Onwards they trudged. Khali cast her eye over the dozens of russet-brown calves that stood out among the herd. These young animals could not fail to draw attention to themselves, dwarfed as they were by the adults. But with such formidable mothers to defend them, and an army of buffaloes to back them up, Khali knew that there was no chance of making a kill.

It was obvious the moment the buffaloes caught the scent of lion. As one, they stopped and stood with their large wet nostrils thrust into the air, swinging their massive heads from side to side as they tried to pinpoint Khali's position. Some snorted and swung around as the whole herd worked itself into a frenzy. Grunting and bellowing, they thundered towards the fallen tree, enraged by the scent of the lioness and her cubs. Khali cowered deeper beneath the tree, tucking her legs under her, desperately trying to avoid the lethal horns of the buffaloes. They pounded the trunk, the bulls smashing at it with the heavy bosses of their horns, trampling the ground into a quagmire. At one point part of the tree rose like a feather into the air, tossed high above the heads of the herd. For the moment, Khali was trapped, hissing and growling, battered by the buffaloes. They hooked at her, trying to dislodge her, until finally, as the black wall of animals closed around her, she wriggled from under their feet and ran for her life towards the forest with half a dozen cows hard on her heels.

But that was not the end of it. It went on and on, methodical and unrelenting: mob violence at its most basic and uncompromising. Another group of buffaloes stampeded towards the place where Khali's cubs lay, eager to join the fray, drawn by the familiar sounds of battle. I kept thinking of the calves that I had seen killed, the long drawn-out suffering of an old bull, his testicles and tail chewed off, mauled to death early one morning by the Marsh lions, and reflected that nature's balance had neither cruelty nor compassion.

Finally, it was over, and the buffaloes retreated. Khali crept back through the forest, a pale wraith drifting among the shadows. The tree lay in pieces, broken and tossed aside as casually as if it had been matchwood. Khali called to her cubs, soft *auus*, repeated over and over, urging them to appear from the rubble. Surely they must be dead? Nothing could have survived such an assault. But Khali was not as convinced as I was. She dug and pawed among the massive pieces of timber, pausing every so often to sniff the ground. Suddenly she bent forward, uttering a low groan of greeting, that I could barely hear. There, confused and bewildered, its eyes pale blue and hazy, sat a single tiny cub. Khali's great pink tongue rocked it from side to side, enveloping it with her scent, reassuring it of her presence. With the darkness separating us, I drove quietly away, wondering what the morning would bring.

Next day, I found Khali resting beneath another fallen tree. I could hardly believe my eyes; all four cubs had survived. The massive root-stock was like a toppled fortress, impregnable even to the buffaloes.

But in the end it made no difference. Confined as she was to one small part of her range, with no males to secure her a place to raise her cubs, and without the Marsh pride to support her, Khali's attempts at motherhood were doomed to failure. As the Bila Shaka pride tightened their stranglehold on the marsh over the next few days, Khali disappeared with her cubs and I never saw them again. Like the rest of her pride she simply vanished.

The young cheetah sat like a spotted sphinx, surveying the marsh. The bulge of her powerful chest muscles made her rounded head look small. Dark tear marks like runny mascara reached from the inner corners of her wide-set eyes to the blackened edges of her lips, accentuating her gaunt look. Her throat was snow white, with no hint of the distinctive row of blotches that hang like a thick black necklace around a leopard's throat.

It was six months since the female and her two brothers had become independent from their mother. During the year and a half that they had depended on her they had always travelled as a group, never straying far from each other. Then one day, mother and cubs split up – just like that, no ceremony, no backward glances. For the next few months the three youngsters hunted together, though it was the female who made the majority of the kills. Now she hunted alone. Her brothers had gone their own way and would stay together for life. A single male rarely has much success in breeding; by forming a coalition, he is better able to acquire and defend a territory.

The young female was the embodiment of feline grace and beauty – as sparsely constructed as the plains themselves. Fine-boned and of slender frame, a cheetah looks almost too delicate to be exposed to the harshness of Africa's wild places. Such thoughts were quickly tempered by the memory of three cheetah brothers who used to roam the acacia thickets bordering the marsh. I had once seen them assaulting a solitary male they had caught wandering through their territory. They hunted him down, growling and biting, raking him with their claws. His legs and hind-quarters were a mass of open wounds by the time the brothers abandoned the wretched creature. Two days later drivers found his half-eaten carcass beneath a seething mass of griffon vultures.

When cheetah cubs become independent from their mother, the males stay together. Females roam over large areas, following the widely ranging Thomson's gazelles. Males try to defend a smaller area where a number of females' ranges converge – but competition is understandably fierce and single male cheetahs often have less reproductive success.

The female walked to a termite mound and stood poised above the blond grass, the white tip of her long, spotted tail flicking restlessly from side to side. In the distance she could see vultures squabbling over the remains of an old kill. If she had been a lion or a leopard perhaps she would have drawn closer to investigate. But a solitary cheetah is not a fighter or a bully; it is not physically equipped to compete with other predators. The female rarely if ever scavenged and could ill afford to feast on her kills in the leisurely fashion adopted by lions. And she lacked the phenomenal strength and retractile claws that enable a leopard to drag the remains of its kill into a tree.

When three vehicles pulled into view, the cheetah ignored them. Already this morning she had been visited by a dozen cars filled with tourists delighted to see yet another Mara cheetah. She had long ago learned to distinguish between the noise of people in vehicles and the sounds of Masai herdsmen and their cattle. Whenever she saw a Masai approaching, she would lower her head, flattening herself against the ground, ready to hurry away if she was

ABOVE: *An almost fully grown cheetah cub chasing a hyaena – though if there had been food, the hyaena would have stolen it.*

OPPOSITE: *In the Serengeti, eighty per cent of cheetah cubs perish during the first six weeks and many more fail to reach maturity.*

detected or chased. But the Masai had no quarrel with the cheetahs. It was the stock-raiding lions and hyaenas that they worried about.

One of the three cars edged closer, only stopping when the cheetah seemed about to leave her perch. A man leaned forward and raised the barrel of a gun, bracing his forearm against the window of the car. Holding his breath, he aimed at the cheetah's shoulder, adjusting for the distance and the breeze. He could feel his own heart pounding with the tension. The gun went off with a crack and the cheetah leaped forward, running a few paces before turning in surprise and retracing her steps. She searched for the cause of the momentary sharp pain that she had felt as the drug-laden dart pierced her skin. It had felt just like the bite of a tsetse fly, except that this was not tsetse fly country. Tsetses sought the shade of the acacia thickets to the north and east, favoured by the leopards.

Without the benefit of colour vision to guide her, the cheetah failed to pick out the bright red plume identifying the syringe, which lay spent on the ground, a few metres from the termite mound. It was barely a minute since the drug had discharged into her shoulder muscle, but already her back legs felt light and weightless; she was sleepy and strangely unconcerned. The drug had been tested many times before and showed no dangerous side effects. A few hours from now the female would have no recollection of her brush with science and would be searching for prey.

Everything was going according to plan. Dr Pieter Kat from the genetics section of the National Museum in Nairobi was in charge of the darting operation. Once he was sure that the cheetah was adequately sedated, he stepped from his vehicle and beckoned to the other members of the team of scientists to begin collecting blood and skin samples from the cheetah. And so began an elaborate routine that had been repeated countless times in the Mara and the Serengeti, in an attempt to learn more about a species that at the turn of the century numbered tens of thousands, and whose range extended through Africa, the Middle East and India. Today cheetahs throughout Africa are so inbred that they are virtually genetic twins, classified by the International Union for the Conservation of Nature as an endangered species.

For reasons that are yet to be discovered, the world's cheetah population went into rapid decline 10,000 years ago. Only a few individuals survived, causing the species to become highly inbred with greatly reduced genetic variability – the raw material on which natural selection acts. Inbreeding also increases the likelihood of inherited disorders and infertility. Male cheetahs have very low sperm counts – one tenth that of domestic cats – with seventy per cent of the sperm showing some form of abnormality. Compounding this, all cheetahs show much the same degree of susceptibility to bacteria and viruses, making it difficult for the population to adapt to an outbreak of infectious disease. Pieter Kat and his team are working on a disease-management plan for areas where the cheetah is thought to be especially vulnerable. As the cheetah's future is known to be at risk, the study of diseases which may affect it is particularly important.

The cheetah faces an uphill battle as it struggles to adapt to a rapidly changing environment. Within the next fifty years, much of Africa will have been settled by the expanding human population and transformed for agricultural purposes. Denied their natural prey, cheetahs have no alternative but to turn to hunting livestock, a habit that has led to them being classed as vermin in Namibia, where they are shot or captured.

Even *within* protected areas cheetahs encounter hard times. A recent study by Karen Laurenson in the Serengeti (harbouring a stable population of 500 adult cheetahs), found that eighty per cent of cheetah cubs died within seven or eight weeks of birth, while confined at a den – often little more than the depression created by their mother's body at the base of long grass or tucked away in a reed-bed. The majority of deaths were attributed to lions and hyaenas, with leopards, smaller cats, large mongooses, baboons and birds of prey known to take their toll. When the cubs are mobile enough to follow their mother, they are still liable to be killed by predators. Karen found that forty per cent of those that survived life at the den perished during the next month.

A cheetah mother will attempt to defend her cubs by bluff and threat, spitting and lunging, even at times slapping or biting as she desperately tries to distract larger and more dangerous predators, while her cubs flee. Fires are another common danger that can search out a litter hidden in long grass, and sometimes a cheetah is forced to desert tiny cubs due to the migratory nature of the Thomson's gazelles, her primary prey.

The cheetah's battle to raise cubs successfully among high densities of large predators has been highlighted by the decrease in lions and hyaenas living in the ranchlands bordering the Mara Reserve. There has been a particularly dramatic decline in the numbers of spotted hyaenas. It is possible that these adaptable predators have simply deferred to the large number of Masai living in the area, keeping a low profile during daylight hours or moving further south, seeking protection within the reserve. One certainly doesn't see or hear hyaenas as often as one used to. However, the more likely cause for their decline is poisoning by pastoralists. Hyaenas do at times attack livestock in the Mara, and the Masai have in the past used Coopertox (cattle dip) or some other form of poison to eliminate predators.

With fewer predators to compete with, the cheetahs are flourishing outside the reserve. A

Cheetahs hunt mainly during the daytime, when there is less chance of conflict with lions and hyaenas. Tour drivers, anxious to please their clients by showing them predators, sometimes follow cheetahs when they are trying to hunt or crowd a female with young cubs. If a mother is disturbed she may try to move her cubs from one safe place to another, making them vulnerable to attack.

number of cheetah mothers have been seen with litters of four or five cubs in tow. But there are other worries. Of all the big cats living in the Mara, the cheetah is the most vulnerable to disturbance by tourism – particularly when denning or accompanied by young cubs. When the grass is long, the drivers are tempted to edge closer and closer so that their visitors can see clearly. Not long ago one of the safari guides told me of a particularly disturbing incident involving a cheetah mother with four two-week-old cubs, who was forced to move her litter due to vehicle harassment. As she carried one of the cubs in her mouth she was trailed by a procession of vehicles. She had gone perhaps twenty-five metres when she put the cub down, its eyes barely open. She looked distressed and agitated, not knowing which way to turn next. Hurrying now, she ran back to the den and picked up a second cub as tawny eagles circled overhead. But this time she moved off in a slightly different direction, dropping the cub as vehicles once more followed her. As she went back to where the first cub lay in the grass a secretary bird ran in and struck out at the helpless creature, stomping on it. For a moment it was a scene of total pandemonium. Vehicles charged towards the secretary bird and chased it back into the sky. Beside themselves with rage and remorse for bearing witness to such a scene, the visitors asked my friend to drive away. When they returned next day they found the cheetah mother sitting in the long grass, with one cub nearby. The others had perished.

—4—

The Wolves of Africa

The hunting dog is one of nature's underdogs. Instead of ruling the predatory kingdom with their prowess, they appear to have been brutally penalized for it, and, with a helping hand from man, the species seems to be sliding slowly backwards towards the dark abyss of extinction.

SERENGETI: A KINGDOM OF PREDATORS
GEORGE SCHALLER

The radio message came across loud and clear at Kichwa Tembo Camp.

We have received word of a pack of wild dogs on your side of the border in the Mara Triangle. Can you go and look for them, and if possible try and photograph them for us – it is the Border Rovers.

THE VOICE WAS THAT OF DR MARKUS BORNER of the Frankfurt Zoological Society, one of the dozen or so scientists based at the Serengeti Wildlife Research Centre. Markus co-ordinated the Serengeti wild dog project, helping to track pack members from the air by using radio collars.

Wild dogs – the most endangered of all the large carnivores. It was months since I had last seen a pack of these enigmatic predators racing across the Mara plains in pursuit of a gazelle or wildebeest, and more than two years since I last enjoyed the company of the Border Rovers. This pack roamed the area astride the border separating Kenya's Masai Mara from Serengeti National Park in Tanzania, and was of particular interest to me.

In 1988 I had spent a number of months in the Western Corridor of the Serengeti watching the wild dogs known as the Ndoha pack, during which time four young females had

OPPOSITE: *The scientific name for the African wild dog is* Lycaon pictus – *'painted wolf'. Each dog has a unique pattern of dark brown, yellow and white markings. The wild dog evolved from the wolf family some six million years ago and is the only canid with four (rather than five) toes on its front feet – it does not have a dew claw.*

A mother wildebeest will fiercely defend her calf against attack by a single wild dog or hyaena. However, when these predators hunt in packs they are invariably able to force the mother eventually to abandon her calf. Thomson's gazelles are the wild dogs' primary prey in the Serengeti-Mara. During the migration the dogs feed almost exclusively on wildebeest calves.

emigrated from the group. Female emigration is a feature of wild dog social behaviour: all young females leave the pack in which they are born in their second or third year to try and form a new breeding pack with unrelated males. This helps to avoid inbreeding.

Throughout the following year the Ndoha females had wandered the length and breadth of the Serengeti–Mara ecosystem, from their natal range in the west near the shores of Lake Victoria to the Gol Mountains in the east. Eventually they headed north to the wooded grasslands of the Mara, covering a total area of 10,000 square kilometres in their search for mates. But so thinly distributed are the dogs that nowhere could they find suitable males with whom to breed; and in the course of their wanderings one of the females disappeared.

Sometime in the latter half of 1989, the Ndoha females established a union with five male dogs, whose origins were unknown. Thus the Border Rovers were founded. The eight adults had been seen from the air with a large number of puppies towards the end of 1989, at a den near the Ngiro-Are anti-poaching camp in the western corner of the reserve, an area known as the Mara Triangle. It was difficult to count them, but there were perhaps twenty puppies, which meant that more than one of the Ndoha females had given birth at about the same time. Sadly, the excitement of finding the pack with puppies was short-lived. A week later the only sign of the pack was a dead adult in the vicinity of the den. Had disease yet again cast its long

Wild dogs feed remarkably amicably at a kill, competing with one another by bolting their food rather than by fighting as lions do. They hunt in the morning and evening, chasing across the plains and wooded grasslands at speeds of up to sixty kilometres an hour. If they have puppies, they will return to the den after they have killed and regurgitate food for their young.

shadow over the future of the wild dogs, a species that is today even more endangered than the black rhino?

That was the last I heard of the Border Rovers until Markus radio-called on 29 November 1990. I quickly sketched a map of where the dogs had been seen, but as was so often the case, the pack had disappeared by the time I found their resting place. Having spent months at a time in the company of wild dogs in the Serengeti, I knew how suddenly a pack might decide to move, regardless of the time of day or how suitable their resting place appeared to the human eye. Wild dogs are born wanderers, trotting away across the horizon with that easy stride of theirs, putting six or seven kilometres between you and them without even pausing for breath. In the Serengeti, a single pack wanders over a home range of approximately 1,500 square kilometres – an area the size of the Mara Reserve. In the Mara, where good concentrations of prey are available year round, a pack can survive in a range of half that size.

Having cursed the dogs' ability to vanish into thin air, I finally found them scattered among the rocks, the dappled colours of their coats blending perfectly with the blacks and browns of the lava slabs. I photographed as many members of the pack as possible, so that each dog could be identified in the future, even if some emigrated to form new packs. By keeping an identification file, scientists can chart the life history of each individual, recording births and deaths, emigrations and relatedness among pack members. Only by being able to

ABOVE: *Wild dogs and hyaenas are coursers, not stalkers like the cats; if necessary they can sustain a chase for several kilometres.*
OPPOSITE: *Wild dogs enjoy resting near water after hunting, wallowing in muddy pools to cool themselves while the yearlings and puppies cavort in the water. Dogs drink whenever water is available, but can go without for long periods.*

identify individual dogs positively is it possible to learn more about such a highly social and far-ranging species.

As I sat enjoying the company of the Border Rovers a car approached. It was Kevin Pilgrim, one of the balloon pilots based at Mara Serena Lodge. Much to my surprise, Kevin told me that he had been visiting the Border Rovers on a daily basis for the last twelve days, ever since he had discovered them hunting in the area on 18 November. At that time he had counted nineteen dogs, twelve of which were puppies. Kevin had never seen wild dogs before. He was fascinated by their extraordinary bat ears and motley coat markings, seduced by the sight of a pack of wild dogs accompanied by a large litter of puppies. Each day, after the morning balloon flight, Kevin searched for the dogs. He noted the way the whole pack cared for its puppies, which were mobile enough to follow the adults as they hunted each morning and evening. Having made a kill, the adults would stand aside as soon as the youngsters caught up, allowing them to eat their fill while keeping the ever-present hyaenas at bay.

For the next month the Border Rovers confined themselves to the open plains and scattered woodlands to the west of Serena Lodge, where there were large concentrations of wildebeest, gazelles and impalas – all species which are highly favoured as prey by wild dogs.

ABOVE: *Thomson's gazelles often flee from predators using a highly visible and stylized gait known as stotting or pronking. This is thought to signal that the gazelle is fit and healthy and that there is no point in the predator giving chase now that it has been sighted. It certainly has the effect of alerting other animals to the approach of a predator.*

LEFT: *All the adults in a pack protect the puppies, although normally only the dominant female gives birth. The dogs are aggressive in defending the den site where their puppies are born against much larger predators such as lions and hyaenas; a pack of dogs can often chase off a group of hyaenas, although they must sometimes yield to greater numbers at a kill.*

Whenever possible the pack singled out the youngest wildebeest among the herds. Each morning, at around 7 a.m., as the two Serena balloons drifted over the plains towards the Serengeti, Kevin scanned the country below him for signs of the dogs. Once or twice he saw the wildebeest scattering in the distance, the dust kicked up by their hooves rising into the air to meet him. There, in the wake of the herd, would be a line of dogs, ears back, white-tipped tails held ramrod stiff, racing after the wildebeest at sixty kilometres per hour. Momentarily shutting off the gas supply which keeps the balloon aloft, Kevin would feel an eerie silence settle over the scene, broken only by the faint sound of the dogs yipping with excitement as they closed on the heels of their prey. He would watch them deftly cutting out a cow and calf amid the dust and confusion; in less than a minute the pack would be feeding on their victim, banishing the calf's mother from the vicinity of the kill.

One of the Border Rovers' favourite resting places after the morning hunt was a shallow pool of water in a murram pit dug to repair the roads. Like wild dogs everywhere, the Border Rovers loved the chance to lie up close to water, cavorting in and around the pool in a mad helter-skelter of activity, then lying at the muddy edges, basking in the heat before seeking shade beneath a nearby fig tree. Their presence at the pond was the cause of great indignation to the resident pair of Egyptian geese. The geese viewed the murram pit as sacrosanct, voicing their alarm at the rowdiness of the pack with nasal honks and quacks that echoed across the plains whenever the dogs were in residence.

On this particular day the pack was within 200 metres of where Kevin had first sighted it, though the collared female had disappeared shortly afterwards, leaving eighteen dogs. There were five adult males and one female. The twelve puppies were about five months old, so they had been born in July or August. This was typical of wild dogs in the Mara. Packs generally give birth to their puppies while the migration is in the northern part of its range. This allows them to take full advantage of the super-abundance of food represented by the wildebeest and their five-month-old calves, while anchored at a den.

The Border Rovers looked in excellent health, particularly the puppies, who had beautiful thick coats richly marked with patches of dark brown, orange and white. The puppies all had very similar markings and were the same size, so were obviously from one litter. My only worry was the collared male, who was violently sick at one point, and walked with a pronounced limp. I looked at the puppies playing together, so full of life, their tummies filled with meat. How many, I wondered, would still be alive a year from now?

By the time I next saw the Border Rovers, both collared dogs had disappeared, leaving us with no reliable way of relocating them. In recent months rabies had decimated a number of packs in the Serengeti–Mara region. So it was doubly important to find the Border Rovers again and inoculate each dog with a vaccine that would help protect the pack. A scientist with the Loita wild dog project happened to be staying at Mara River Camp at the time, but he was unable to find the dogs before returning to Nairobi.

By the New Year there was still an abundance of prey in the Serena area, though most of the migratory wildebeest had already headed back to the Serengeti plains, some 200 kilometres to the south. This forced the Border Rovers to switch their attentions to the large herds of impalas and Thomson's gazelles which had grown sleek and fat on the lush grazing provided by the short rains. Though it was still occasionally sighted, the pack never stayed long enough in one place for scientists based in Nairobi to fit a radio collar to one of the dogs. My only hope was that the pack might den in the Mara during July or August, when the plains and woodlands of the Mara Triangle would once again reverberate to the sound of the massed armies of the wildebeest migration.

Wild dogs have had a chequered history in the Mara. Historically they have fared poorly within the reserve, where the densities of the dominant predators – lions and hyaenas – are high. Nevertheless, every so often word would go round among the drivers that wild dogs had been sighted again. But they never settled; they were always moving, seldom staying in any one part of their huge home range for more than a day or so. Only when it had puppies did a pack abandon its nomadic existence, confining itself to a den for the three months necessary to ensure that the young could safely follow.

Wild dogs often den somewhere that has served them well in the past. One such place was Aitong, an area of gently rolling plains some thirty kilometres to the east of Mara River Camp. Here, among the Masai pastoralists, lions and hyaenas lead an almost exclusively nocturnal existence to avoid conflict with wandering tribesmen. This is to the dogs' (and the cheetahs') advantage. They are primarily diurnal hunters, chasing after their prey in the early morning and late evening, while the lions and hyaenas are forced to remain hidden.

The plains surrounding Aitong Hill are speckled with patches of acacia bush and croton thickets. It was here in 1977 that Joseph Rotich introduced me to Black Dog and the Female, the dominant breeding pair of the Aitong pack. I would hardly describe our meeting as love at first sight; they and their relatives were a motley bunch. Except for a thick ruff around his neck, Black Dog was completely hairless. As for the Female, she was a perfect caricature of one of those milk-laden bitches engraved on Greek motifs. But once I learned to recognize them as individuals, I became immersed in the social world of the pack and its daily struggle for survival. Black Dog, the Female and three other adults were all that remained after a disease-ridden year had laid waste the rest of the pack and its twelve puppies.

Black Dog and the Female finally perished in 1982, having somehow withstood the intermittent impact of epidemic diseases such as distemper and rabies. For the next three years wild dogs were absent from the wooded grasslands surrounding Aitong Hill. Then, in 1985, out of nowhere, a pack of nine dogs appeared and a new era unfolded. No one could remember seeing a pack quite like this before. Between 1986 and 1989 the new Aitong pack thrived, producing two litters of puppies each year. At one point in 1987 their number swelled to more than forty individuals, giving substance to the stories of bygone years, when large packs were said to be common in many parts of Africa. The dominant female and her sister gave birth at the same time, without unduly stressing the pack. This was contrary to what has so often been witnessed in the Serengeti, where only the dominant female normally manages to raise a litter. When another female attempts to breed, the dominant female usually tries to kill her puppies or recruit them as her own. Perhaps it was the year-round abundance of food in the Mara that enabled the Aitong sisters to raise two litters simultaneously. With each successive year, some of the previous generations of puppies emigrated, usually as two-year-olds, wandering off in single-sex groups to search for mates and establish new packs, helping to replenish the dwindling wild dog population of Mara and Serengeti.

Then, in late August 1989, disaster struck. Suddenly dogs started dying at Aitong, just as they had ten years earlier. The dominant female was once again ensconced at a den with a new litter of puppies. None survived. Within the space of a few weeks the pack was practically wiped out. By the time I left the Mara in November, more than twenty adult members of the pack had died of rabies. Miraculously, two of the yearling males survived and later joined up with two unknown females from the Lemek area to the east. Perhaps they would breed in the coming year, to continue the run of success enjoyed by their relatives. But it was not to be; within the year they, too, had died. I was utterly despondent. What hope was there for the survival of these beguiling animals? Everything seemed to be against them.

But there was good news from elsewhere in the Mara. In June 1989, a small pack denned near Mara Intrepids Camp on the east side of the reserve, not far from the Talek River, producing a litter of eight puppies. The Intrepids pack had been created by the union of two

The dominant female in a pack gives birth once a year. If a subordinate female succeeds in giving birth, the dominant female may try to kill or adopt the puppies. Because the normally nomadic dogs are tied to a den for three months while puppies are young, it would strain their resources if there was more than one litter to feed, or if puppies were born more than once a year.

male dogs from the Serengeti with two females from the Mara region. The adult males, one of whom wore a radio collar and was therefore known as Collar, were brothers from the Ndoha pack in the Western Corridor. It was their sisters, the Ndoha females, who had emigrated a year earlier and eventually helped form the Border Rovers. Collar had left his natal pack towards the end of 1988 and was later joined by a younger brother. The two Ndoha males had really travelled. From the palm-fringed banks of the Mbalageti River to the spotted land of Mara involved a journey of almost 200 kilometres. But their search for mates had not been in vain.

When their puppies were four months old, the Intrepids pack pushed north, filling the vacuum created by the demise of the Aitong dogs. During the middle of 1990 Collar's younger brother departed and was seen shortly afterwards in the company of two unknown females, both of whom were beautifully marked with yellow flanks. The urge to breed for himself had overridden the younger male's social loyalties to Collar and the five surviving Intrepids puppies.

In mid-June, the young male and one of the two females produced a litter of puppies. They were now known as the Ole Sere pack – it means 'goodbye' in Masai. Their den was situated to the east of the road leading from Aitong town to the Talek gate, a few kilometres beyond the reserve boundary. Here, among the Masai and their cattle, there was plenty of food:

wildebeest from the Loita Plains, topi, zebras, and numerous impalas and gazelles. But there were no trees to offer shelter on these rolling plains – only hour after hour of searing heat. How I wished I was a wild dog so that I might escape from the sun in the long grass or crawl into the shade of a darkened burrow. Instead I would just have to wait until the day cooled and the dogs became active; it was forty kilometres back to camp.

A herd of fifty cattle grazed within view of the den. In their midst was a young herdboy, perhaps ten years old, spear braced behind his neck, razor-sharp blade glinting in the sun. He was dwarfed by the rangy cattle, only his head and the tops of his shoulders visible over their backs. The mother of the puppies was anxious at the closeness of the boy and his cattle. She stood motionless, staring out across the plain, never taking her eyes off the herdboy who wandered around the periphery of his cattle, curious as to all the attention the dogs were generating among visitors from the tented camps.

Eventually the male dog rose to his feet and stretched, then ran towards the female. Round and round they went in a frenzy of greeting, peeing and licking each other's faces. Then they trotted over to a bare mound of earth at the edge of a clump of long grass. The female paused, nervous of bringing her puppies above ground. But eventually the adults poked their heads into the darkened entrance of the den, backing away as four tiny brown and white puppies emerged, barely five weeks old. The female flopped on her side to let them suckle, shielding them from view within the mouth of the burrow.

Each time people went out to visit the new pack there were fewer puppies to be seen. At first there had been six; then only four emerged, then three, then none. The cause of their death was unknown. When I next saw the Ole Sere pack towards the end of the year, there were five dogs – the two adults had been joined by three yearlings, two males and a female. This was a surprise to everyone and nobody was quite sure where the young dogs had come from. But their presence was obviously welcome: for the moment, five dogs were more of a pack than two.

In September 1990, I was told that seven wild dogs had denned to the east of Aitong Hill and given birth to ten puppies. It was the Intrepids pack. By mid-November, their puppies were big enough to abandon the den, allowing the pack to travel more widely. I finally caught up with them one evening to the south of the main road between Leopard Gorge and Mara Buffalo Rocks, places that I often visited in my constant search for leopards. All the adults had lost their hair due to sarcoptic mange, and their naked grey skin contrasted sharply with the brightly coloured coats of the twelve-week-old puppies.

What a pleasure it was to see the dogs again. But three of the puppies had by now disappeared, although the seven survivors appeared to be in good health. Like all wild puppies, they were full of life, chasing and playing, pulling and pushing, generally annoying the adults. As for the five yearlings, they relished the chance to play with their younger brothers and sisters, inciting them to follow or to chase. The pups had just shared a gazelle fawn brought back by their father. But still they were not satisfied, pestering the yearlings to regurgitate some of the freshly consumed meat from an adult male Thomson's gazelle they had killed.

The Intrepids pack were last seen alive in December by members of the Loita wild dog project. The research workers were mounting a campaign to try and protect the wild dogs from the ravages of epidemic diseases, particularly rabies. Each Masai manyatta in the area had been visited by the team and the owners of domestic dogs asked if they would agree to their being inoculated. Interestingly, not one of the 200 people censused said that wild dogs killed their stock. The Masai keep a close eye on their herds during the day time, preventing them from becoming scattered and vulnerable to predation. In late January 1991, John Richardson, the veterinary surgeon with the wild dog project, located a radio collar worn by a member of the Intrepids pack, during one of the regular aerial searches. It had been cut through with a sharp instrument, then abandoned. Perhaps someone had found the dead dog, removed the collar and been too frightened to keep it. John had hoped to track down the Intrepids pack, all of whom had been vaccinated against rabies, so that he could check antibody levels. But they have never been seen again.

Meanwhile the Ole Sere pack had also fallen victim to disease. Two of the five dogs were found dead by rangers, who reported the incident to John Richardson. Fortunately, John was able to fly to the Mara and collect one of the carcasses, which showed very high levels of rabies virus in its saliva. The chances were that the other members of the pack would also have been fatally infected, and sadly this proved to be the case.

George Schaller's words quoted at the top of this chapter ring as true today as they did thirty years ago. Little has changed for 'nature's underdogs'. The wild dog is one of the world's most endangered canids, with perhaps fewer than 3,000 individuals remaining in the wild. Formerly found in thirty-four countries south of the Sahara, the dogs are threatened with extinction. Currently the only significant numbers are in Kenya, Tanzania, Zambia, Zimbabwe, Botswana and the Kruger Park in South Africa. The dominant female in a wild dog pack breeds each year, producing an average of ten puppies – sometimes twelve or fourteen, even sixteen in one instance in Serengeti. So, in theory, a pack can reproduce itself annually. But the wild dogs seem particularly vulnerable to epidemic diseases such as distemper and rabies, which can wipe out whole packs within days. Parvo-virus, anthrax, Babesia, canine ehrlichiosis and sarcoptic mange are also known to be a threat to the dogs.

But man's prejudices are still probably the most potent weapon aimed at the dogs. Despite efforts to influence public opinion and portray the wild dogs in a more positive light, many ranchers still reach for their guns at the very mention of them. And few, if any, wildlife sanctuaries are large enough to protect viable breeding populations of such widely dispersed predators. Sadly, Africa's 'painted wolves' are running out of space, hemmed in on all sides

ABOVE: *Wild dogs always select the youngest or weakest animal in a herd of wildebeest – calves, yearlings, pregnant cows or, occasionally, old and vulnerable males. A wildebeest will sometimes seek refuge from a pack by backing up against a vehicle – in which case the driver should quickly move away and allow nature to take its course.*

LEFT: *The presence of the migration supplies the wild dogs with ample prey during the dry season in the Mara. Wild dogs only weigh about twenty to twenty-five kilogrammes, but by hunting in a pack and running their prey to the point of exhaustion, they are sometimes able to kill adult wildebeest or zebras who may be ten times that weight.*

by the human population explosion and the increasing threat of diseases contracted from domestic animals.

Despite the dramatic losses incurred by the wild dog population, some have always survived to disperse and form new packs. For years there might be no word of the dogs in an area. Then one day they trot over the horizon and lay claim to the open plains as if they have never been away. Would the Border Rovers return later in the year? With the loss of the Intrepids and Ole Sere packs to the east of the Mara River, the Border Rovers were the only pack of dogs still thought to be alive within the vicinity of the reserve.

—5—

The Old One

The darkness of the earth at night is never complete in Africa, because even the darkest night sky has a glow of light behind it.

GOING HOME
DORIS LESSING

OVER THE YEARS, I had become resigned to a fleeting relationship with predators such as wild dogs and leopards; their presence was like a breath of fresh air, their restless wandering adding vitality to the Mara year. But I never grew to rely on being able to find them, was never sure from one day to the next if or when I would see them again – that was part of their fascination.

The lions were different. They were a consistent feature of the Mara, confined by the limits of individual pride territories. So it was in the company of lions that I spent the majority of my waking hours. Whenever I failed to find lions around the marsh, I headed south, beyond Rhino Ridge, searching for the Paradise lionesses, or by chance crossed paths with members of the Gorge pride while searching for leopards. As for the Bila Shaka lionesses and their cubs – they could, of course, be found 'without doubt'.

During the past few months I had spent most of my time in the Mara Triangle, glassing the plains in front of camp each morning for signs of the Kichwa Tembo pride, before continuing on my way in pursuit of the Border Rovers. Whenever I was fortunate enough to locate the wild dogs, I found myself comparing their behaviour with that of lion society. Lions are certainly not as social as wild dogs, a fact revealed by the level of aggression shown by pride

OPPOSITE: *The Old One. Lionesses eventually stop breeding and rarely kill for themselves. They remain within their territory, though the rest of the pride may become less tolerant of them, particularly when food is scarce.*

OVERLEAF: *Some of the 1.6 million migratory wildebeest begin to arrive in the Mara in June or July. By this time the Serengeti Plains are dry – there are no permanent rivers there – and the wildebeest must seek a more plentiful supply of grass.*

members at a kill. Serious fights over food among members of a wild dog pack are unheard of: they compete for their share of the spoils by the speed with which they eat, rather than by fighting. Having hunted, wild dogs regurgitate food for any individuals who are ill or who have stayed at the den to protect the puppies. If a dog is separated from the rest of the pack, the others invariably attempt to find it; calling and searching, calling and searching, until the pack is reunited amid a frenzy of greeting.

So what did the future hold for lionesses like the Old One of the Kichwa Tembo pride? Do members of a pride 'help' the old or the ill? I wanted to discover whether or not these old lionesses were allowed to live out their lives peacefully within the pride or whether they were penalized for their age once they no longer bred and were unable to hunt for themselves.

I remembered years earlier seeing an old and pathetically thin lioness approach the well-eaten carcass of a buffalo that the Bila Shaka lions had been feeding on. The pride had recently undergone a take-over by new males and only now, after two or three months of disruption and hostility, was life beginning to return to some sort of normality.

The younger females in the pride had quickly come into oestrus and mated with the new males. But the old lioness was no longer fertile. Consequently, the new pride males took little interest in her; she was barren and rarely made a kill that they could feed from. The old lioness had become an outsider within her own pride territory. As she moved towards the remains of the dead buffalo, one of the males got up. His belly was laden with meat, but he was still not prepared to tolerate another lion feeding at 'his' kill. He threatened the female, lunging forward and driving her away from the carcass. But the old lioness was now desperate with hunger and before long she once more attempted to draw near and feed. Again the male rose and strutted towards her, and this time he clubbed her to the ground with a flurry of massive blows from his forepaws. The lioness toppled sideways and lay still. The male stood there, towering over her, his mouth pulled back in an aggressive snarl, the enormous shadow cast by his body seeming to obliterate her presence. It was a sickeningly brutal sight, with no semblance of pity, forcing me to reconcile the harsh perspective of wild lions. Was this, I wondered, the fate awaiting the Old One?

One morning during the dry season of 1990, I found all eleven members of the Kichwa Tembo pride resting together about a kilometre from camp; they had long since returned from their foray over the escarpment. The Old One sat among her relatives, the hair on her nose sparse and grey, her nostrils black and spotted. When she yawned I could see that one of her lower canines had been snapped off and worn almost down to the gum, and the other was reduced to little more than a stump. The old lioness seemed to be shrinking before my eyes. Each day she looked smaller, less robust; she was spending most of her time sleeping. There were days when I thought she must be dead and I repeatedly caught myself glancing at her chest to see if she was still breathing.

By contrast the younger lions in the pride looked sturdy, their heavily muscled limbs quite unlike those of the elderly lioness, whose rump and thighs had caved in. The Old One never stirred when the five youngsters suddenly sat bolt upright, responding to the soft *auus* of their mothers. Her battered ears barely flickered in acknowledgement of the sound. The cubs stared out across the ash-blond plain, their eyes riveted on a column of wildebeest moving single file through the long grass. By now the migration was among us again and the Kichwa Tembo pride territory was alive with the grunting and lowing of tens of thousands of animals.

Some evenings the pride killed two or three wildebeest as the herds gathered to drink at the Mara River. The sudden arrival of the wildebeest had been like an Indian summer for the Old One, her life flourishing briefly with the glut of food. There were many times when I

Old lions find it difficult to compete at kills with younger, stronger members of the pride and may eventually lose condition and become emaciated. Lionesses in the Serengeti-Mara are known to have lived for seventeen years, and one in Nairobi National Park is recorded as having reached twenty.

found her out on the open plain feeding, long after the rest of the pride had retreated to their favourite resting place within a tight circle of croton bushes overlooking the river. She would gnaw at the bare ribs of the wildebeest with her blunt molars, licking the last scraps of meat from the carcass as the sun beat down on her ragged coat.

But her rejuvenation was short-lived. When I next saw her in early August 1990, she had sustained the most horrific injury to her back right foot. The skin and muscle had been torn away, leaving the white tendons protruding like porcupine quills from her paw. But to my surprise she managed to walk without too much of a limp, though every so often she stopped and gently licked her injured paw. To add to her misery she bore two fresh fang marks just above her hips – the result of a confrontation with her son Blond Mane at a kill. The strapping male knew that he could bully his mother and often behaved belligerently towards her. He had little respect for any creature weaker than himself. Perhaps it was he who had injured her foot.

ABOVE AND OPPOSITE: *Only by crossing the Mara River can the migratory wildebeest enter the area known as the Mara Triangle, where even during the height of the dry season there is usually grazing for the herds. They return every year to favoured river-crossing sites where the more experienced individuals have crossed before, often using narrow hippo trails as exit points.*

The following morning I found the Old One lying by herself, coughing and wheezing in the long grass. Unable to settle, she hauled herself out into the open, continually twisting round to try and lick her back. But she could not reach the suppurating wounds and clean them. The five cubs watched the old lioness, reacting with curiosity at her ponderous movements, confused by her presence. They seemed almost wary of her, as if she did not quite belong. They encircled her, sniffing at her backside, trying to provoke some kind of response. But eventually they left her in peace and sought the shade.

The majority of the wildebeest meanwhile had moved further east, crossing the shallow sections of the river with ease. Now the herds were massed on the open plains surrounding Aitong Hill, where the wild dogs used to roam. Smoke from numerous fires set by the Masai to rid themselves of areas of unnutritious sward filled the air, blurring all colour to a uniform dullness. The abundant rainfall had yielded a surfeit of grass, allowing the resident plains game to thrive in even greater numbers. Topi and Coke's hartebeest, impalas and Thomson's gazelles had all flourished. But some time in the future, drought would return to thin out the wild herds and the domestic ones alike.

The Kichwa Tembo pride's territory was now virtually empty of game. Only small parties of gazelles, impalas, zebras and the occasional thin black line of wildebeest clustered on the burnt areas, which had turned green in the wake of recent showers. The lions dozed fitfully,

As long as the river is shallow, comparatively few wildebeest perish, although crocodiles and lions may still lie in wait for them. The zebras are often the first of the migratory herds to arrive in an area, proceeding in a much more orderly fashion when crossing the river, and it is practically unknown for one of them to drown.

their hunger prompting them to keep a constant watch for prey. As they slept, the weasel-like faces of five banded mongooses poked cautiously from the air vents piercing the top of a nearby termite mound. The mounds are a common feature of the Mara: miniature earthen castles dotted across the sweep of the plains. Some are two metres high, elevating a lion or cheetah well above the remnants of the prey-concealing grass.

The sudden arrival of the Kichwa Tembo pride earlier in the morning had forced the banded mongooses to keep quietly hidden. They had smelt the pungent cat odour, as one of the lions had flopped down on the mound, snuffing out the windows of light. The year-and-a-half-old lioness knew they were trapped beneath her, but did not bother to investigate. Her playfulness had begun to mellow these past few months, now that she was of an age where she could begin to try and hunt for herself. As a young cub she had relished the opportunity to chase and kill the many smaller animals that shared her world, toying with creatures such as mongooses as she refined her fledgling hunting skills. Like all cats, lions are born with an array of instinctive behaviour patterns, such as the ability to stalk and ambush, the tendency to crouch and pounce. But there is a world of difference between what a young lion might think is suitable prey and what would cause a more experienced pride member to hunt. Young lions have to learn which types of animals are rewarding to pursue and whether or not it is worth investing time and energy in attempting a kill. Most important of all, they

ABOVE: *This five-month-old cub is attempting to suffocate a heavily pregnant wildebeest captured by one of its older female relatives. Lions have an instinctive ability to stalk, pounce and bite, but do not become adept hunters until they are one and a half to two years old. Until that time they rely for food on kills made by the adult members of the pride.*

LEFT: *Cheetah mothers often catch and then release gazelle fawns, allowing young cubs to practise their fledgling hunting skills. Cheetahs become independent at fourteen to eighteen months, but even at that age they are not very proficient hunters.*

must learn the patience to wait – and wait – until the moment is right to launch an attack.

Eventually the young lioness rose to her feet and stretched, pausing an instant to sniff casually at one of the air vents before joining her four brothers in the shade of the bushes nearby. Feeling the vibrations of her departure, the mongooses began to stir deep within the maze of passageways. A few minutes later the boldest of the troop popped its head out of a vent and cautiously looked around. Once it was certain that the danger had passed, it scurried down the steep side of the mound and began to ferret around the bases of the grass stems. Shortly afterwards, it was joined by half a dozen members of the twenty-strong band. Others kept a wary eye on the lions, churring and whistling whenever one of the big cats got up or turned towards them. Despite the mongooses' legendary reputation as killers of snakes, the great majority of their diet consists of insects, supplemented by birds' eggs, nestlings, lizards and small rodents. Searching for food out on the open plains is always dangerous; but moving around in large groups helps protect them from predators.

At times there is an almost telepathic sense of communication between members of a pride of lions. They know each other so well through years of close association that they can sense the slightest change in another's mood or posture. Suddenly every lion within the Kichwa Tembo pride was focused on the warthog heading straight towards them. Squint, one of the Old One's daughters, strode quickly and smoothly through the grass to intercept the warthog, which was blindly unaware of the lions. Blond Mane followed her lead, veering off at right angles, creating a pincer movement. All that was needed now was for one more lioness to creep into position and the warthog would be trapped within a rapidly tightening noose of predators. But just as it seemed certain that this would happen, the warthog stopped as suddenly as if he had run into an invisible wall. His head jerked upwards, his large disc-like snout sniffing the air, and as I glanced at the grass stems swaying in the breeze I knew that the warthog's keen sense of smell had saved him.

Visitors in their mini-buses chattered excitedly among themselves, astonished that the lions had not pressed home their attack and run the pig down. But adult lions know when to abandon a hunt and conserve their energy for more fruitful efforts. That vital element of surprise had been lost. The warthog might not have known exactly where the lions were, but by then the adrenaline was surging through his body – he was ready to flee. Warthogs possess a surprising burst of speed – usually enough to keep them one step ahead of the lions and leopards. However, this wily pig proceeded with exemplary caution as he distanced himself from that dangerous smell. Rather than gallop for his life and risk blundering headlong into the vice-like grip of another lion, he trotted off in the opposite direction to live another day.

The Old One had seen nothing of this. She had not stirred when Squint and Blond Mane had moved out to try and ambush the pig. Even if she had, she would still have been too slow to corner the warthog and would have been excluded from sharing such a small kill. The most she might have hoped for would have been to scavenge from any scraps left by the others. Instead she sprawled some twenty metres away. But she did not rest in the manner of the other lions. She lay so flat among the grass – rested so totally – that the alertness that is such a part of the way lions 'catnap' was missing. Lying like this, she could easily have been run over by one of the vehicles manoeuvring for position

Hippos kill more people than either buffaloes or elephants. When a hippo is on land, never get between it and water – they will charge if threatened; their long, curved canines can cut a man in two with one bite. Lions occasionally kill hippos when they venture on to the plains to feed at night, but a healthy adult is unlikely to succumb to such an attack.

around the pride or trampled by an irate herd of buffalo. In fact, a few days earlier she had narrowly escaped death in the massive jaws of a bull hippo. The bull had taken to abandoning the safety of the river before dark and was making his ponderous way across the Kichwa Tembo Plain. He almost stumbled on top of the Old One, who had not heard him approach. Startled, the hippo charged the lioness, his curved ivory canines chomping with fury as he tossed his mallet-shaped head from side to side. Somehow the Old One managed to slither out of the way, avoiding the bludgeoning swipes of the hippo's jaws, as he careered past her and continued on his journey.

The constant greetings, the sense of doing something together – whether walking side by side or lying in ambush – now played little part in the Old One's daily routine. Her life was no longer clearly attuned to the rhythm that joins pride members; with each new day the gulf between her and the other lions grew wider. The benefits of the Old One's years of experience, gleaned from a lifetime in the Kichwa Tembo area, had already been passed down from one generation to another. Knowledge of seasonal hunting grounds, safe places in which to lie up or give birth to cubs – all this had been handed on. Now her presence only strained the tempers of her pride mates, particularly whenever food was short.

Blond Mane was particularly quick to dominate the Old One, staring at her in a way that forced the lioness to avert her gaze for fear of provoking an attack. It was as if she no longer gave out the right signals. Sometimes he would crouch, ready to charge as his mother moved slowly towards the place where the rest of the pride was lying, or when she suddenly raised her head out of the long grass. Even if there was no kill to fight over he might cause her to grovel submissively at his feet.

The light was fading fast now. The first *auu* was like a whisper – it was barely audible. Only when the other lions added their voices did the Old One finally look up. She peered through the curtain of grass, staring at the place where the five cubs had gathered. They yowled and rumbled in greeting as they moved towards Blond Mane, rubbing up alongside him. The Old One began to groom, then let out an enormous sneeze. She rolled on to her belly and sat crouched over her great paws, combing the long hair on her chest with languorous licks of her rough tongue. As the breeze stirred the grass, she thrust her nose into the cool air, drawing in the familiar scent of her relatives. Squint had already walked away towards the Sabaringo lugga, groaning softly as she departed, urging the others to follow. A small party of hadada ibis flew low overhead, calling, *ha-da-da, ha-da-da.* Abruptly their raucous cries were drowned by the deafening sounds of Squint roaring.

At once the Old One's bearing changed. She surged out of the grass, seeming to grow in stature, thrusting her head and neck forward. Now she too began to call. Her roars were rough and menacing, and the power of her voice startled me – so much life and energy pulsing from deep within that wasted body. Suddenly she was one of them, full of vigour. Her body shuddered with the effort, head alert, eyes filled with a new brightness. She moved out one more step, then lay down facing the direction that Squint had taken. The urge to follow had already dimmed and the moment passed.

A few minutes later Squint called again and this time Black Beard and Blond Mane added roars of their own – longer, deeper roars that distinguished the pride males from the lionesses, and would serve as a warning to other lions that this territory was spoken for: 'We are here, beware.' The five young lions stayed silent. They would not begin to roar properly until they were about two years old, so it was even more essential that they should maintain close physical contact with the adult members of the pride on whom they still depended for food and protection.

The Old One walked slowly towards her other daughter and sniffed beneath her tail. In the last of the light I could just make out the distant shape of Squint standing in the centre of the plain, staring back over her shoulder, waiting for the others to catch up. The Old One struggled to join her, walking with slow, laboured paces. The cubs merged, pressed one against another, caught up in a world of chasing and play-fighting. Black Beard and Blond Mane stared after the lionesses and cubs, but made no attempt to follow. Their lives were lived by different rules. They would catch up in their own good time. The males would have no difficulty tracking the lionesses by their roars or their scent, joining them when hungry or if one of the females was in oestrus. Until then, they would continue to patrol the pride territory, marking the fallen trees and bushes with claw and urine. The lionesses roared again, the sound echoing back and forth, as each one added her voice, clearing an invisible path ahead. Then the darkness swallowed them up.

That was the last I saw of Black Beard and Blond Mane. No one knew for sure what had happened to them. All male lions show an inherent urge to explore beyond the bounds of their pride territory. The only restraint on encroaching further afield is the presence of

ABOVE: *Termite mounds are an ideal resting place for lions and cheetahs, enabling them to keep an eye out for prey, especially during the rainy season when the long grass impedes visibility. These Kichwa Tembo lions, a female on the left and two males, are about eighteen months old – note the beginnings of manes on the males.*

OPPOSITE: *Adult male lions gain much of their food by scavenging from the kills of pride lionesses or other predators such as hyaenas, cheetahs and wild dogs. Once they are too old to defend a territory they are forced to become nomadic again, and may incur the wrath of the Masai pastoralists by trying to kill livestock.*

resident pride males in the adjoining territories. Males – be they nomads or pride males – constantly probe for any weakness in the established order, seeking the chance to extend their influence. Perhaps the Kichwa Tembo males had found other, unprotected females, and were mating? They had sometimes been seen loitering across the river, taking advantage of the vacuum created by the disappearance of the Marsh lions. But when I searched the Musiara area I found only the Bila Shaka pride, who by now had completely forsaken the Miti Mbili lugga and were firmly ensconced in the marsh.

Had the Kichwa Tembo males taken one of their not infrequent journeys up the Isuria Escarpment after the departure of the wildebeest from the Kichwa Tembo Plains? The lions could always find plenty of resident game among the wooded grasslands on top of the escarpment. In fact, pockets of wildebeest and zebras often retreated up the steep hillside just

before nightfall at the height of the migration. One rarely saw the pride when they travelled beyond the reserve boundary. They acted quite differently, keeping well hidden during the daytime. Here, among the acacia thickets and Combretum woodlands of the escarpment, there was a new danger.

The escarpment signalled a transition in cultures. Beyond the rocky knife edge is poachers' country: patches of forest interspersed with open plains, a land spotted with thorn-bush and in places beset with the wire snares of the Luo, Kipsigis and WaKuria. Each year tonnes of game meat are butchered and set out to dry, before being swifted away to the surrounding villages. What was once a sustainable form of subsistence hunting is being commercialized by greedy middlemen.

There is no love lost between poachers and predators. Lions and hyaenas are always willing to play the part of scavengers, taking advantage of an easy meal held in a wire noose, particularly at night, when the poachers are safe in their camps. Consequently, predators are usually the first animals to disappear from an area of heavy poaching. The poachers use guns and poison, sometimes putting a wire snare at the only entrance to their thorn-bush hideaway among the denser thickets, trapping any predator incautious enough to try and steal from their larder of drying meat.

Cheap to buy and endlessly reusable, snares kill and maim unselectively. I have seen an elephant cow with her trunk sliced in half; a bone-thin lioness with a necklace of wire pulled so tight that it cut a hole through her throat.

There were other dangers. Black Beard and Blond Mane would have been fully aware of the possibility of conflict with the Masai if they trespassed: the escarpment was crisscrossed with cattle trails. But there was no guarantee that a hungry lion might not try and kill a cow or calf, incurring the wrath of the herdsmen. On one occasion I had seen the Kichwa Tembo lionesses hurrying back down the rocky face of the escarpment. They looked nervous, as if anxious to shelter within the invisible protection of the reserve, their faces bloodied from a recent kill. Had it been livestock, perhaps? If so, the Masai would not easily forgive or forget. Some time in the future a lion would pay with its life for such audacity. No full-blooded Masai youth would walk away from the chance to capture a lion mane bonnet, thus bringing pride and honour to his generation of age mates.

Whatever the reason for their disappearance, it was hard to accept that Black Beard and Blond Mane might really be dead. The Kichwa Tembo drivers had watched Black Beard every day for the past four years, observing with interest as Blond Mane rose to pre-eminence within the pride. Increasingly the younger male had been able to dominate his father, the irrepressible vigour of youth giving Blond Mane the edge whenever the two males fought for possession of a kill.

Without the reassuring presence of Black Beard and Blond Mane, the lionesses and their cubs lived with a gathering sense of unease. The effectiveness of the roars from the pride had been diminished. Soon the Out of Africa pride realized that there were no longer any males defending the adjacent territory. At first they were cautious whenever they trespassed, hesitant lest they should suddenly find themselves confronted by their enemies at close quarters. But as time passed, the two old males became bolder, resting openly during the daytime along the Sabaringo lugga which lay at the heart of the Kichwa Tembo pride territory. Proof enough, I reasoned, that Black Beard and Blond Mane were gone for good.

One morning during this period, a new driver at Kichwa Tembo Camp told me that he had just seen a lioness who was so weak that he feared she might be dying. The lioness had multiple injuries, some of which were fresh. One side of her body was a mass of deep wounds. As the driver moved closer the lioness dragged herself into the shade of a croton thicket. He felt certain that this must be the lioness we called the Old One.

When I reached the place he had described to me there was no sign of the old lioness,

ABOVE: *The Masai treasure cattle above all other possessions. A man's status – and his ability to purchase a wife – depends on the size of his herds. Masai culture is based on a hierarchy of age-sets and each generation of boys must go through a series of traditional ceremonies in order to become a warrior and eventually an elder of the tribe.*

OVERLEAF: *By co-operating with each other, lionesses are capable of killing animals far larger than themselves – a single lioness is rarely successful in pulling down a full-grown buffalo, which may weigh up to 700 kilogrammes. In this instance, it was the added weight and strength of the Marsh pride male which tipped the balance in the lions' favour.*

although three other females lay resting nearby. I immediately recognized them as members of the Out of Africa pride. They must have found the Old One as she wandered in search of her pride. I imagined the scene; the angry grunts and snarls, the Old One slow and confused, barely understanding what was going on, throwing herself on to her back in order to appease the fury of these other lionesses, desperate to escape. But lions can be very unforgiving at times. The old lioness – whoever she was – had been beaten to within an inch of her life.

Imagine, then, my joy a day or so later when, having feared the worst, I met the Old One and her two daughters walking side by side with two of the young males. Their stomachs were laden with meat: I had never seen the Old One look so replete. Whatever the kill had been – in all probability it was a buffalo – it had provided ample food for the whole pride. I looked carefully at each of the lions, searching for the tell-tale signs of conflict: fresh wounds to face or flank, a nervousness that might reveal a change in the balance

ABOVE: *Lions, like all predators, are opportunists, always on the look-out for sick or injured animals. The young of many prey species are kept hidden away from the rest of the herd, or have formidable mothers to help defend them. This young buffalo had a broken leg which prevented it from keeping up with the herd and provided the lionesses with an easy kill.*

OPPOSITE: *A group of elephant cows and calves can defend itself against any number of hyaenas and lions, but if a calf becomes separated it is doomed. The mother-calf bond is very strong – a calf continues to suckle, folding its trunk back over its head and using its mouth to drink, until it is four years old and its mother gives birth again. If a calf dies, its mother will stay with the body for days.*

of relationships within the pride or might hint at mayhem with the Out of Africa pride.

Eventually the Kichwa Tembo lions reached the sparse shade of a Gardenia tree, its crown a mass of white, trumpet-shaped blossoms. The Old One was first to arrive, flopping to the ground beneath the tree. One of her daughters pushed her head against her mother's chin, rumbling in affectionate greeting. The young males hurried to join them. They jostled one another for position, rubbing their heads against the broad face of the old lioness, caught up in a moment of intense friendliness. Watching, only a few metres away, crouched an African hare. Its large brown eyes bulged on the sides of its head as it lay with long ears flattened protectively along its back. It had frozen motionless when it heard the swish of grass as the lions swept through, the mottled greys and browns of its coat helping to camouflage it in the long grass. Now it was tensed like a coiled spring, ready to bound away if seen.

I waited with the pride until nightfall, watching them greet again as the light faded. All of the lions were now on their feet. They peed: the cubs squatting, the females scraping their hind feet, leaving an unmistakable mark as they raked the urine-soaked earth. Elephants drifted by, making not a sound. The younger lions wanted contact, flopping down in front of their mothers. But the lionesses ignored them - watching, waiting, expectant. They knew that it was time to be on their way. Safari vehicles hurried back to camp, the sounds of hippos groaning and snorting in the river ringing in the visitors' ears. The light had almost gone and a chorus of frogs began to build: *tink, tink, tink*. Rufous-naped larks echoed a final call, their song floating back across the plains from every bush and termite mound.

Gradually, in twos and threes, the lions drifted away across the plain. I stayed with the Old One, barely able to make her out in the growing gloom. Her ears were cocked as if she were listening, and every so often she looked around as if expecting to find the other lions still resting nearby. If only she had been able to turn her head further she would have seen them: seven dark lion shapes moving flank to flank in a broad front, only visible to my eyes for as long as they were silhouetted against the horizon. But by the time the Old One stood up, they had vanished into the night. I waited, thinking that she might roar, but she did not. It was as if she no longer really cared: all that mattered was that she was here within the invisible bounds of her pride's territory.

Shortly after this, in mid-January 1991, the balloon pilots at Little Governor's Camp told me that they had seen the Old One lying at the edge of the swamp in front of the camp. 'She looked in a terrible state,' said one of the pilots. 'She kept on retching.' Prior to this I had found the Old One resting up in a patch of forest just beyond the fence surrounding Kichwa Tembo, and on one occasion she had stumbled on a waterbuck calf lying motionless in the bushes. But that had been brief respite indeed. For the rest of the time she had survived by scavenging on the withered carcasses long since abandoned by the hyaenas and vultures, loitering within sight of a Masai manyatta on a slope overlooking camp. Perhaps the herdsmen had put out poison.

'The Old One is dead.' Those were the words that greeted me on my return from a brief visit to Nairobi. The news filled me with a sense of great sorrow, tinged with relief. At last it was over.

It had happened during the second visit that the Old One had made to the Masai manyatta. How many other times must she have been tempted during the past few months, but not dared to act? She had come at night, squeezing her thin body through the thorn-bush enclosure, clawing her way into a roofless wooden hut used as a temporary holding pen for sheep and goats. The Masai had heard the wild bleating of frightened animals and knew that a predator was among them. They came carrying torches, picking out the silvery green reflection from the half-blind eyes of the old lioness as she scrambled out with a sheep in her jaws. Half a dozen other carcasses lay scattered on the dung-spattered floor of the hut; robbed by domestication of a means to escape, they had been easy prey. By now the Old One had apparently abandoned all attempts at finding the rest of her pride. All she could think of was her hunger. Desperate for more food, she crept back the following night, an hour after dark.

The Masai knew the lioness would return. Their long-bladed spears – taller than a man and oiled so that they would pass cleanly through skin and flesh – rested close to their beds of stretched cowhide. The two men slept lightly. Their senses were attuned to the slightest sound. Any hint of uneasiness among their cattle would wake them in an instant.

And so it was. Suddenly they were wide awake. Not a word was spoken as they rose from their resting places. They got silently to their feet, each with a spear clasped loosely in his right hand. They crept to the entrance of their cow-dung houses, waiting for their eyes to adjust to the darkness. A sliver of moonlight illuminated the boma. The fear of the cattle hung like a pall over their dwelling, poisoning the cool night air. The men could see the animals pressed tightly against one side of the thorn-bush enclosure. A few metres away, the Old One struggled to snuff the life from a calf, scarcely able to contain her desire to start feeding.

Her senses dulled with age, the Old One probably never heard the men approach. Perhaps her hunger was too great, the taste of warm blood and meat too much for her. Suddenly, she found herself trapped in a bright beam of torchlight. Barely understanding the danger, she froze over the calf. Two spears sliced through the stillness and thumped into her chest, piercing her heart. She hardly moved. A long drawn-out rumble shook her weary frame. Then she pitched forward to sprawl in an untidy heap across her last meal.

Next morning, the Masai reported the incident to the game rangers stationed at the Oloololo gate, two kilometres away. There would be no charges made against them. They were simply exercising their legal right to protect themselves, and their livestock, from the deadly threat of a large predator. Later the men dragged the Old One unceremoniously from their manyatta. Young children jostled one another for a better view of the predator that had dared to kill in their midst. The men hauled her to the bottom of the hill and dumped her in the long grass. The Old One had been their enemy. She had broken the rules. Now she had paid for her transgression.

By the time I found the Old One a few days later, little remained of her body. I pushed aside the grass covering her skull, examining the worn and broken teeth, the tattered patchwork of her skin, marvelling at the way the old lioness had endured for so long. I wondered if the rest of the pride would search for her. Somehow I doubted it. Her presence had become so fragile that her fate would be of little consequence to them.

But even as the Old One faded into a memory, another life was unfolding, full of youth and vigour. I was about to become involved in the concerns of a solitary cat with a spotted coat; a cat entirely dependent on its own resources, that had to remain hidden if it was to survive among the lions and hyaenas.

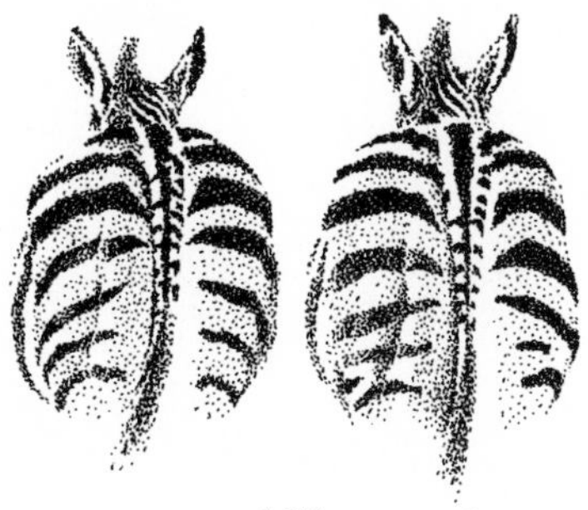

—6—
The Paradise Female

Lions are animals of the sun, leopard seem to be of the moon. Though both hunt by night, the lion has always been associated with myths of the sun; he is shining and radiant; his mane frames him like the sun's rays. The leopard's coat is like the dapple of firelight and dark on the forest floor, his eyes are the pale gold of the hunter's moon.

A Glimpse of Eden
Evelyn Ames

It is just like old times. An early rise long before dawn, the rock-strewn road jarring the sleep from my body. I watch as the sky turns pink, then crimson, fading again to mute, cold colours until the sun breaks the horizon. My spirits soar, the drowsiness fading as I reach the familiar corner just before the old Mara Bridge where, each morning, white-browed robin chats serenade the dawn.

I turn off the murram road on to the faint winding track, passing recently burned acacia scrub. Everywhere looks sparkling fresh, with the green of new shoots piercing the fire-blackened topsoil. Four kilometres later I stop and scan the rocky outline of Fig Tree Ridge with my binoculars. A slight movement – a rustling of leaves in the top of a solitary fig tree – catches my eye. The leopard is still there, draped extravagantly along a thick branch, her kill hanging beside her. In the distance I can see other vehicles racing towards the ridge.

Scraps of meat and fragments of bone litter the ground beneath the ancient tree; clusters of white fur plucked from the belly skin of a young Thomson's gazelle dance like feathers on the wind, pasted together by the leopard's saliva. Nothing will go to waste. A column of safari ants are swarming over one of the scraps of meat. Mongooses, genets, honey badgers and birds of prey – a myriad of animal life will feast here later, if the hyaenas do not get there first.

OPPOSITE: *The Paradise female. A leopard may have a number of favourite trees within its home range which it regularly uses as resting places.*

OVERLEAF: *Dawn and dusk are dangerous times for prey animals, as predators tend to rest during the heat of the day and hunt when it is cooler. The predators' specialized nocturnal vision also gives them an advantage over the prey species.*

Though there are more leopards than cheetahs in the Mara, cheetahs hunt by day and are therefore much easier to find. Leopards lie up during the daytime, often up trees or in rocky outcrops or thickets. Their stealth and use of camouflage help them to avoid contact with their enemies and launch surprise attacks on their prey.

I listen as the leopard shears through skin and sinew with her sharp carnassial teeth; followed by a rhythmical *crrrrunch, crrrrunch* as her molars pulverize the gazelle's slender leg bones. The pupils of her eyes are dilated, black holes letting in the light. This is her second kill in two days. Last evening I found her feeding on another gazelle carcass in this same tree. She looks replete, but is hurrying to eat as much as possible, as if anxious to slip away before many more cars arrive.

Now begins the fastidious process of cleaning her face and paws. Balanced on three legs, she rhythmically licks her long pink tongue along the inner edge of her forepaw, deftly wiping it across the side of her face; cleaning from her nose and lips the blood and tiny scraps of meat that would otherwise attract a horde of flies to her face. Her ivory claws glint in the sun as she flexes a massive forepaw and meticulously licks them clean.

Suddenly she is on her way down, all fluid grace and suppleness, pausing in the crotch of the tree to look around her. A ripple of excitement flows through the watching crowd. This is the moment we have all been waiting for. I hold my breath, willing her not to slip from view down the back of the trunk. Her spotted coat fills the viewfinder of my camera. Is there anything more beautiful than a creature such as this, golden and shining in the soft early morning light? It is the briefest of moments; a cursory glance to left and right and then she is gone, leaping the final two metres to earth, cushioning her descent with the thick pads of her forepaws. The show is over. The visitors settle back down in their seats, happy just to be able to say they have seen her, and the vehicles depart.

The leopard's solitary lifestyle and secretive habits endow it with an unmistakable aura. It is considered by many to be the most strikingly beautiful of all Africa's animals. But when I first visited the Mara it was virtually impossible to find a leopard. They were viewed as a stock-raiding menace by landowners, who on occasion put out baits laced with cattle dip: a cheap and deadly way of ridding themselves of predators. But the greatest threat to Africa's leopard population was the price to be earned from that spotted skin. Those that survived the poaching onslaught of the 1960s and '70s became fugitives, masters of stealth and concealment. When they hunted, it was invariably during the hours of darkness.

In those days you could spend a month in the Mara and the only trace of leopard you might see would be a blur of spots disappearing through the long grass. Yet even that proved sufficient to enrich the evening hours spent discussing the day's game viewing around a camp fire. Nowhere was safe from the poachers. The Serengeti's famed Seronera Valley, where the sight of at least one leopard was virtually assured during a safari, struggled to live up to its reputation as the best place in East Africa to watch leopards.

By the early 1970s public opinion had begun to change. It was no longer fashionable to wear spotted cat-skin coats, and the International Fur Trade Federation called for a voluntary halt to trade in the more endangered species. In 1975 the snow leopard, clouded leopard, tiger, jaguar, leopard and cheetah were listed on Appendix 1 of CITES (the Convention on International Trade in Endangered Species of Wild Fauna and Flora) – a treaty that has since been adopted by 109 countries – prohibiting international commercial trade in these species.

Some African countries still allow leopards to be hunted as trophies, though export and import licences must be obtained in order to move the skin out of the country of origin. Today, most conservationists agree that it is loss of suitable habitat, rather than poaching, that poses the greatest threat to spotted cats (leopards, jaguars and tigers are also still shot by landowners as a real or supposed threat to livestock or humans). But there is no doubt that significant trade in spotted cat skins – particularly of Latin America's smaller cats – continues, despite international agreements. It could be the deciding factor pushing some of them to extinction.

I had almost given up hope of studying and photographing leopards when in 1978 Joseph Rotich told me that a very shy female had given birth to two cubs in Leopard Gorge, a short game drive away from Mara River Camp, where I was based. Though I never managed to get close enough to identify their mother, I did on occasion succeed in finding the two cubs. The male cub was much shyer than his sister. Chui, as I named her, was always different – the most approachable of leopards living in the area I was familiar with – and over the years I came to recognize her and to learn something of her ways.

Four years later, towards the end of 1982, Chui's mother gave birth to a male and a female cub at Mara Buffalo Rocks, a massive stone fortress with a network of caves running through its heart. Within six months Chui, who by now was five years old, produced her second litter of cubs at the Cub Caves, one of her favourite haunts along Fig Tree Ridge. Both mother and daughter had chosen wisely. Under less arduous conditions the females would have moved their cubs from time to time, to avoid attracting the attention of predators. But drought had robbed the Mara of much of its cover: the ground was bare, bushes trampled and stripped of concealing vegetation. Even at night, moving around with cubs was just too dangerous. So the leopards stayed – and stayed – attracting visitors from all over the world. For anyone wishing to see a leopard with cubs, now was their chance. Soon everyone was asking for directions to Fig Tree Ridge and Mara Buffalo Rocks.

In early 1984, three months after I had first seen them at the Cub Caves, Chui and her cubs suddenly disappeared. By then, Light and Dark, as the cubs were known, were six months old and big enough to follow Chui safely wherever she pleased to wander. For days I searched throughout the Northern Ranchlands, but could find no trace of them. Surely someone, somewhere must have word of Chui? But always I was greeted by the drivers with a shake of the head, or an upturned palm. As the months passed, I became resigned to the fact that Chui had gone, merging once more into the shadowy world that helped her to secure prey and protected her from her enemies. Like the countless other leopards that are never seen, she had proved that she could survive from one year to the next without anyone being the wiser – just as her mother, the Mara Buffalo female, had done for so many years. Freed from Light and Dark's dependence on the Cub Caves, Chui had slipped back into that intriguing world which makes the leopard such a sought-after animal.

Then, nearly two years later, in October 1985, friends told me they had seen a rather shy female leopard in Leopard Gorge. Film-makers Warren and Genny Garst had only glimpsed her through binoculars, but felt certain that she had cubs. Her belly skin was loose, they said, and the hair around her four swollen teats was brown and matted. Sure enough, the following day I found three leopard cubs playing among the rocks and bushes surrounding a cave on the north face of the gorge. In fact it was within metres of the place where Joseph had told me that Chui and her brother had been born. At one point during the afternoon the cubs' mother melted into view. She looked nervous, peering cautiously from behind bushes. She snarled and hissed in the direction of my vehicle, warning me to keep my distance, even though I was a hundred metres away, separated by the wide mouth of the gorge.

It was impossible to identify the leopard positively from such a distance, but I did manage to take a single photograph of her with my 800mm lens before she disappeared. When I returned the following day both she and her cubs had vanished, though Warren Garst found her again a few days later. By this time she had moved to the Cub Caves on Fig Tree Ridge. She now seemed less wary of vehicles, though she quickly ushered the cubs into one of the caves when two of the Gorge pride suddenly emerged further along the ridge. That was the last the Garsts saw of the female and her cubs, and the following day I departed for England.

Nearly a year later, I happened upon the leopard's picture while sorting through a mountain of slides. Knowing that each leopard has a unique pattern of spots, I set to work trying to identify her. Having checked the female's facial markings with photographs of a number of leopards I had encountered over the years, I picked up a copy of my book, *The Leopard's Tale*, and flicked through the pages. Something caught my eye. I stopped and went back to a picture I had taken of Chui at Leopard Gorge in late 1981, the time when she had given birth to her first litter of cubs, one of whom had been killed by a lioness. When I compared the pattern of markings of the two leopards, they were identical. It was Chui. So she *was* still alive. How I regretted not having been able to take a closer look at her, to have rediscovered something of her familiar presence. I would have searched for fresh scars gained in spats with other female leopards, checked her ears which would have become worn and ragged over time, seen how much darker her pink nose had become. I longed for just one more look at Chui, to absorb the feeling of an older, wiser leopard.

That was five years ago. If Chui was alive today, she would be thirteen years old, an age that few wild leopards attain. She would certainly be nearing the end of her life, and in all probability was dead.

ABOVE: *This photograph of Chui and one of her cubs, Dark, on Fig Tree Ridge was taken in 1983. Cubs are weaned when they are three months old, though they may continue to try to suckle until they are five months and their mother will start bringing small kills such as warthog piglets or gazelle fawns back to them when they are only four to six weeks old.*

RIGHT: *Leopards begin climbing when they are six to eight weeks old, and by three months they may begin to follow their mother to where she has stored a kill in the safety of a tree. Their extraordinary tree-climbing skills often enable them to escape when surprised by lions and hyaenas.*

Leopards mark their home range by spraying their powerful scent on to fallen trees and bushes. This will indicate to another leopard that the area is occupied; or attract a wandering male to a female in oestrus. All leopards will seek out such scent posts to find out more about what is going on in their range.

Now I found myself in the company of another female who seemed to promise me the chance to observe her habits. One of the drivers told me that the Paradise female had been born three years earlier, to the south of Governor's Camp in the Paradise area, hence her name. Once she had reached full independence at one and three-quarter years of age she had left her mother and moved north from Paradise Plain. Apparently the drivers had seen her at various times during the next few months, tarrying for a while along the Miti Mbili lugga.

Regardless of where the Paradise female came from, I had never met a leopard like this before. She was so tolerant of vehicles that she considered them to be a movable – and 'markable' – part of her home range. She would sometimes rub her cheek against the bumper of vehicles, spraying her scent against the wheels, just as Black Beard, the old Kichwa Tembo lion, had done. I have even watched her crawl beneath the back of a truck, using it as a shield while stalking prey.

In February 1990 she had been seen mating with an old male leopard whose territory overlapped her home range. Despite all the years of driving through this area, I had yet to see this particular male, who by all accounts was a huge animal with a thick jowl of fur hanging beneath his chin. Anxiously I awaited the moment when the Paradise female gave birth to her first litter of cubs. Recently she had become more localized in her movements, consistently prowling among the acacia thickets surrounding Leopard Gorge and Fig Tree Ridge. But

A mother leopard moves her cubs periodically to help avoid the unwanted attentions of predators such as lions and hyaenas. From the age of about three months the cubs will start to follow their mother round their home range, though they will usually remain hidden while she is away hunting.

even here there were dangers. While consolidating her claim to a share in the area, she kept a wary eye out for the lionesses of the Gorge pride. They, too, favoured Leopard Gorge as a birthplace for their cubs. And there was always the formidable opposition of the troops of olive baboons that wandered over every part of the leopard's fifteen-square-kilometre home range.

It was just after 11 a.m. when I spotted the green safari vehicles from Governor's Camp. Usually at this time of the morning the drivers were making their way along the steep banks of the Mara River to show people the crocodiles and hippos during their second game drive of the day. But these two vehicles were far from the river, and I knew by the way they were speeding along the track towards Fig Tree Ridge that a leopard must have been seen; nothing is more guaranteed to induce excitement among drivers and visitors. Word must have gone round camp when the drivers returned for breakfast. My first thought was that it must be the Paradise female. But there was a chance that it was another leopard, a shy female whose home range overlapped with that of the Paradise leopard. She had given birth in April 1990 to a male and female cub in the Cub Caves, where Chui had raised Light and Dark. Though the female cub had since disappeared, the male had become quite tolerant of being viewed, unlike his mother, whom I had never seen.

I followed the vehicles' progress with my binoculars. In the distance I could see the glint of sunlight reflected by other safari trucks clustered together at the base of Fig Tree Ridge. People were standing up, craning out of the roof hatches of every vehicle with cameras pressed to their faces. As I approached, one of the drivers held up two fingers, then dipped his hand in imitation of a leopard weaving through the grass. The shy female and her cub were here.

I angled my car so that the cub's mother would not see me until she had fully emerged from the cave. I was desperate for a good look at her, in the unlikely event that it was Chui. The drivers rarely saw the female, and even when they did she invariably retreated to the sanctity of one of the caves or hurried to distance herself from the vehicles.

The stillness was absolute. There was no breeze. The heat pressed in like a great weight on my chest. A sudden movement caught my eye. Was it her? A small pointed face peeked over the rocks, its rusty brown eyes staring intently towards the cave where the leopard and her cub rested. *Grrrrrr, grrrr, grrrr,* it cried, two or three times in quick succession. The slender mongoose could not see the leopards, but he certainly smelled them.

Too late I spotted the female when she was already halfway up the vertical wall of the cave. I cursed myself for having assumed that she would stay hidden and grabbed my binoculars. There was no time to focus a camera on her. At first I thought it must be the cub until she paused for a moment, her head framed against the blue sky, fixing me with piercing yellow eyes. She looked small. This was definitely not a leopard that I recognized from the past. It certainly wasn't Chui. She had been much larger, with a darker coat.

As she reached the low line of acacia bushes on top of the ridge, the tsetse flies swarmed out to meet her. She hurried away, snapping her teeth and lashing her tail as the tsetses' needle-sharp mouth-parts lanced through her loose, spotted coat. Then, in one long, liquid movement, she vanished over the top of the ridge and melted away in the tall grass. I listened to hear if the baboons had spotted her. But there were no barks of alarm. I decided to wait and see if the male cub would appear, but he remained hidden until after dark.

Next morning I could find no sign of the leopards along Fig Tree Ridge, so I continued on my way, hoping to find the Paradise female. Every few hundred metres I switched off the engine and listened, straining my ears to pick up the sounds that might reveal her presence. Finally I heard what I had been hoping for: nervous, yelping alarm calls – jackals barking. I hurried to the spot, pausing every so often to pinpoint their position until at last I could see them moving around with their dainty, stiff-legged gait. They paused to sniff the long, dry grass stems, their keen noses sifting for clues. Perhaps they already knew where she was; knew how long it was since she had passed this way; knew if she had killed and what her victim had been. I could only ask myself such questions.

I stopped to scour the crown of an ancient ebony tree, where I had once found Chui resting along its bare arms, the fresh carcass of a young impala dangling by her side. It was one of the few occasions that I had managed to find Chui with a kill in clear view, away from all the cars. The tree had been one of her favourite resting places. There she was safe from the lions and the hyaenas, hidden from the tsetse flies. Now the tree lay sprawled on its side, its enormous sculpted form resplendent in a thick coat of silvery bark. When it crashed to the ground, it had torn free from the brown earth, uprooting giant slabs of red rock, creating a perfect daytime hideaway for puff adders and porcupines. A few metres above, a harrier hawk scuttled over the gnarled grey limbs, ignoring the spotted hyaena resting in the shadow of the up-ended trunk. Prodding and probing, flapping and floundering, the hawk thrashed its long wings to keep its balance as it searched for food. Giant wood-boring beetles and striped skinks wedged themselves even tighter into the old tree's cracks and crannies, trying to avoid the hawk's long, scaly legs and outstretched talons.

In the distance I could see a tractor grinding through the thickets, scouring the land for firewood. With each passing day the staff from the tented camps were forced to travel further and further afield, searching for wood fuel to service the kitchens and heat the boilers that provided hot showers for visitors. Increasingly, the land is robbed of essential minerals needed to help replenish the soil. Soon the firewood will be finished, unless the camps use less wood-intensive heating systems or convert to solar heating. Perhaps tomorrow Chui's old resting place would be gone for good. Chains would be slung around its outstretched limbs, the tinder-dry wood splintering and cracking as the tractor wrenched the branches free, the hyaenas and porcupines forced to move again.

I drove on, crossing Leopard Lugga, travelling the familiar rocky path that leads towards Leopard Gorge. Even if it were not prime leopard habitat, the gorge would always be worthy of a visit because of the sheer beauty and mystery of the place. I have explored its caves on foot, felt the chill, cool air that attracts the daytime leopard, smelt the heavy ammoniac reek of the hyrax droppings piled centimetres deep on the floor. At times I still come here just as the light is fading from the sky and sit and listen to the night sounds echoing between its rocky walls. It is an eerie spot: a haven for hyrax and eagle owls, agama lizards and Gabon nightjars, the *chack-chack* of a European wheatear mingling with the raucous chatter of a party of Ruppell's long-tailed starlings.

I never fail to look up – half expectantly – at hiding places that would appeal to any leopard, more in hope than the belief that I might find what I am looking for. There are times when I know that all the searching will be in vain. The bush hyrax are relaxed and huddled together, enjoying the early morning sun; baboons scream and squabble over some social transgression as they plunder the fresh crop of figs and ripening fruits from the African greenhearts. Under these circumstances, any leopard in the vicinity would already have crept into the safety of one of the caves or moved on.

It felt wonderful to rediscover this most favoured of places. Little seemed to have changed. Agama lizards scurried across the rocky outcrops. Tsetse flies bit deeply into my skin as I tried to concentrate my vision

ABOVE: *The bateleur* (Terathopius ecaudatus) *is one of Africa's most colourful eagles. It has unusually long wings, enabling it to soar for hours over the plains – the name bateleur comes from the French for tightrope walker or acrobat. It is usually a scavenger, but can kill small game such as dik-dik and mongooses, particularly if it has chicks to feed. The bird on the left of this picture is a male, on the right a female; they are distinguished by the black band along the trailing edge of the wing, which is broader in the male.*

OPPOSITE: *Euphorbia trees are common on rocky hillsides in the Mara and in Leopard Gorge, which lies a little to the north of the reserve boundary. Leopards sometimes store their kills in Euphorbias and baboons break off the tree's fleshy branches and rip off their outer casing to feed on the juicy pulp.*

through binoculars, scanning the caves and crevices where I had spied leopards in the past. A pair of red-rumped swallows flitted back and forth, cementing their mud houses beneath overhanging rocks. Here the passing years seemed but a blink in time: the swallows were building in almost exactly the same position as five years earlier. Patches of vegetation lay flattened by the passage of a variety of animals – topi and impalas, buffaloes and zebras – who from time to time made their way cautiously through gaps in the rocky ridge. A colony of bush hyrax crouched motionless, staring with unblinking eyes into the steel blue sky, watching the long wings of a bateleur eagle cartwheeling overhead. I paused for a moment, wondering if that was all the hyrax had found to stare at. Had the Paradise female been here earlier?

—7—

Man and Predators

The range of the human mind, the scale and depth of the metaphors the mind is capable of manufacturing as it grapples with the universe, stand in stunning contrast to the belief that there is only one reality, which is man's, or worse, that only one culture among the many on earth possesses the truth.

OF WOLVES AND MEN
BARRY HOLSTUN LOPEZ

THE BLACK LIMBS OF ACACIA BUSHES and scattered Balanites trees beyond Leopard Gorge traced abstract shapes against the sky. Lolling comfortably along a wide branch lay the leopard I had been searching for, head tilted at an angle to the warming sun, tail dangling in a loose twist. If it had been any other leopard she would have slipped from her perch above the plains the moment I started towards her. The fact that she did not even turn her head as I approached marked her out as the Paradise female.

Two bright red patches of blood stained the grass at the base of the tree beneath the spot where a Thomson's gazelle hung, two metres above her. To the north, the Paradise female could see Masai cattle confined in a tight huddle within a thorn-bush enclosure. Men stood rubbing their hands together over the dying embers of the fire, red robes pulled over their heads to shut out the chill dawn air. They knew the leopard was there, but as long as she kept clear of their livestock they would not harm her.

Every so often the Paradise female stared out from her tree. Her gaze was still and resolute, piercing in its intensity. Then she closed her eyes. Does any other animal know how to relax quite like this? Her solid, muscular limbs hung loose, her chin stretched far out along the rough bark of the desert date tree whose crown had been neatly pruned underneath by passing giraffes. Every so often her ears flickered as sounds dragged her back from the world of half sleep, her heavy-lidded eyes opening behind the grey nictating membrane, then lazily closing again.

OPPOSITE: *The moon sets over an acacia tree – one of the most evocative sights in all of Africa. A full moon makes hunting easier for nocturnal predators and was traditionally an ideal time for cattle raids among the WaKuria and Masai.*

The Masai rise at dawn – the women to milk the cows, the men to discuss where the cattle will graze for the day. Their traditional way of life revolved around finding enough grazing for their cattle, which they revere, believing that God bequeathed all cattle to them and that they are the rightful owners.

The Paradise female stared past the Masai, alert to the faint sound of another car approaching. It was time to prepare for the transition between day and night, to hide from daytime wanderers such as baboons and man. A silver-backed jackal trotted across the plain towards the tree where the leopard stood, ready to descend. Her bearing changed as she watched it. She hurried to the ground, retreating to a patch of dense croton that perfectly mirrored the site of a long-abandoned manyatta. But the jackal kept coming, lured by the smell of meat, until suddenly it came to an abrupt halt, nose pointed forward. The jackal stiffened in alarm, instinctively stepping sideways, uttering a harsh scolding yelp. How many times in the past had I welcomed that sound, hurrying to trace the source, knowing that it would almost certainly lead me to a leopard?

From 300 metres above the trees, a tawny eagle began its descent, picking out the shape of the gazelle among the leaves. The wind hissed through its long primaries as it parachuted out of the sky. And as it did so, the Paradise leopard charged from the bushes, giving physical presence to the pungent scent that had caused the jackal to falter. She sprang forward, locking her razor-sharp claws into the trunk, then stood panting astride the first fork of the tree, staring up at her kill. The eagle flung wide its wings, stalling sharply, and turned away.

The leopard hesitated only for a moment, then wrenched the carcass from the snagging

The Paradise female rests in a Balanites tree in the early morning. She has spent the night hunting, and will soon seek a better hiding place in which to lie up for the day. Although leopards are now much more visible in the Mara than they were in the 1970s, sunrise and sunset are still the most likely times for spotting them.

thorns. The sight of the eagle was more than she could endure. Now that her kill had been discovered, her only option was to move it.

With the Tommy firmly clasped between her jaws, she began her descent, the full weight of the carcass swinging crazily from her mouth. I thought that she must drop it, for it weighed at least twenty kilogrammes; but using every ounce of her phenomenal strength, she somehow held on. Reaching the ground, she paused to draw breath, making sure that neither lion nor hyaena were on their way to intercept her. Then, straddling the gazelle and grabbing it by the throat, she lifted its forequarters clear of the ground and began to drag it towards the croton bushes. A pair of spotted stone curlews blinked, watching as the leopard passed within a metre of where they stood, their perfectly camouflaged forms hidden among the speckled rocks and leaves.

Later in the afternoon the Paradise female once again deposited her kill in the desert date. But I could find no sign of her among the thicket. I decided to wait, hoping she would return as the sun began to set.

Night came swiftly. Slowly the sounds of Masai voices and the lowing of their cattle faded into the darkness. A wood owl hooted and then fell silent. A dog barked and the leopard turned her head to listen; she relished the flesh of canines, both domestic and wild. The stars shone brightly in the darkened sky. At last the Paradise female began to move towards the

ABOVE AND OPPOSITE: *Leopards are incredibly strong and have immensely powerful limbs and neck muscles. In the Mara they feed mostly on Thomson's gazelles and impalas, and often store their kills in trees. There they can feed on a carcass for several days without fear of losing it to lions or hyaenas.*

tree, her silky outline gliding easily across the ground. She crouched, then leaped forward, gathered herself and bounded up the trunk, shards of bark flying beneath her as she went. I could see her now, silhouetted against the hunter's moon, bowed as if in supplication before starting to feed.

During the night the wind had gusted, causing trees to creak and doors to bang. I had hoped that perhaps it might bring brighter weather and that I would arise to a clear sky. But the day dawned grey and gloomy, and the road-weary drive to where I had left the Paradise female seemed to take forever. When I was within view of the Balanites tree I looked through my binoculars to see if she might be feeding. But the tree looked strangely empty and the kill had gone.

As I drove slowly around the croton thicket I came face to face with the Paradise female. She seemed nervous, her pupils widely dilated. But her belly was full, and it was obvious that she had feasted throughout the night. She was indeed fortunate that there were so few

Hyaenas are perhaps the most adaptable of all predators. They normally live in groups – known as clans – but an individual may also hunt on its own. Despite their reputation as scavengers, hyaenas are capable of killing animals as large as wildebeest or zebras. They have powerful jaws which enable them to crack open bones to feed on the marrow; as a result their droppings are bone white.

hyaenas in this area to challenge her on the ground. Small and compact, she stared across at the Masai and their cattle, her ears twitching and her white tail-tip flicking from side to side.

The Masai herdboys could contain their curiosity no longer, and moved their cattle towards the thicket where the Paradise female stood watching. They drove her in front of them, shielded behind the massed ranks of their herds. The plains exploded with sound, came alive with grunts and snorts, whistles and blasts as the various species of antelope recognized the predator walking in their midst.

The Paradise female looked thoroughly put out at being forced to vacate her resting place and abandon her kill. Ahead of her, a party of helmeted guinea fowl raked their curved claws through the damp soil, clucking and churring as they searched for insects. Seeing the leopard, they raced towards her, like a flock of broody hens, *tut, tut, tut . . . tut, tut, tut'ting* in protest, hurrying her onwards. The jackal was there, too, joined now by its mate, letting out an ear-piercing chorus; and a male impala ran to within twenty metres of the leopard, then stood fast, rigid with alarm, muscles rippling beneath his sleek brown coat.

Little wonder leopards prefer to stay hidden. It is as if they feel naked without somewhere to mask their presence. They are forced into solitude by the reaction they create. The variety of the alarm calls that they induce is a tribute to their great adaptability. A leopard will kill and eat just about anything, including other predators – the young if not the adult. Lion and cheetah cubs, jackals, mongooses, wild cats and civets, are all taken at times.

There are dangers to be avoided, too. Cloaked in a coat of black and gold, leopards do well to merge with their surroundings. Hyaenas are the scourge of all predators smaller than themselves; baboons never miss an opportunity to mob the animal they perceive to be their greatest enemy; and lions will kill them if given the chance.

The Gardenia tree hardly seemed a suitable place for a leopard to rest. It was a squat mass of tightly packed branches, its leaves cropped close by browsing giraffes. But the Paradise female did not hesitate. She crawled up its twisted trunk, squeezing herself into the smallest of spaces between the branches. It was not ideal, but it would do. By the time the jackals caught up with her, she was concealed from even the sharpest vision. No one would dream of looking for her here. Soon her eyes closed, and she drifted off to sleep.

Noses to the ground, the jackals turned and trotted back towards the croton bushes, picking out the leopard's pugmarks as surely as a blind man reading braille. Deftly they traced the path she had taken, stopping to sniff where she had sprayed the trunk of a fallen tree, cocking their legs to drown her smell with their own pungent odour. When they reached the thicket where the leopard had left her kill, they slowed to a nervous walk. Her distinctive scent in such dense cover made them nervous of moving closer. The jackals could see the kill, but dared not enter the bushes. They peered and probed, bobbing their heads in agitation. Carefully they circled the thicket, guided by their acute senses. Finally they found the discarded stomach of the gazelle. The jackals tugged and sniffed at the reeking offal, chewed a while and then abandoned it. They were not really hungry.

Two days later I found the Paradise female walking across a rocky hill to the east of Leopard Gorge, her long tail curved high above her back, revealing the white along the underside of its tip. Suddenly she froze, desperate to avoid yet another of her almost daily confrontations with the troops of olive baboons. She held her ground, crouched snake-like, head pressed across one of her paws, hoping she had not been seen.

But the adult baboon was not to be fooled. He recognized a leopard in any shape or form and barked out a warning. Instantly an audience of seventy other baboons turned their heads to listen. The loudness of his cry, the intensity of his gaze, all told them that this was urgent: a leopard had been discovered. Like a well-trained troop of soldiers, they converged on the sound. The big male baboon needed no other encouragement than the sight of the troop racing to support him. He charged, his long mane bristling around his neck, looking huge in comparison to the Paradise female. She coughed out an explosive warning, but it made no difference. The male was no stranger to bluff and threat. Years of battling for dominance with other male baboons – and other leopards – had hardened his resolve. He charged again, forcing the Paradise female to lunge towards him, teeth bared in a desperate bid to halt his headlong rush. It bought her precious seconds, a moment's grace in which to turn and flee.

As she spun round, the big baboon was joined by another male. They closed in fast behind her, reaching

ABOVE AND OPPOSITE: *It is often said that baboons are among the leopard's favourite prey. This may be true of some parts of Africa where easier prey such as impalas and gazelles is less plentiful, but during the daytime in the Mara, leopards do everything to avoid contact with these intimidating primates.*

out with dexterous fingers to snatch at her rump. Momentarily knocked off balance, she faltered and lost her footing. Feeling their hands upon her, she flipped on to her back to lash out in desperation as they bit at her. Then she was away again, racing through the rocks.

Other members of the troop hurried to intercept her as she neared a clump of bush, and for a moment she was encircled by the mob, lunging this way and that to keep the baboons at bay. A tree was no use to her now. I had seen the way baboons mob a leopard isolated in the top of a tree. They were as agile aloft as she was. Twice the Paradise female ran, and twice she was forced to try and hold her ground in a patch of bush. Each time she was routed by the screaming and barking and the open-mouthed threats. But eventually she made good her escape, streaking into the safety of a lugga just ahead of four male baboons. Here in the dense and darkened undergrowth they would not dare to attack, fearful of the leopard's lightning charge.

Baboons are highly intelligent animals. They understand the risks involved in confronting predators and measure their response according to past experience; whether they are facing a lion, a leopard or a cheetah, they weigh the odds carefully before responding. The vigilance of baboons helps ensure that leopards keep to cover during the daytime. This troop were interested in pursuing the Paradise female only for as long as she remained in the open, where they could see their adversary clearly.

At twenty-five to thirty kilogrammes, a male baboon weighs very little less than a female leopard, and has powerful canines. Baboons have been known to cause leopards serious – even fatal – injuries. Baboons mob leopards whenever they can, signalling to the predator that it has been seen and forcing it to move away to hunt elsewhere.

Search as I might, I could find no sign of the leopard's final retreat. Thinking that she was probably sleeping on the floor of the lugga, I busied myself with writing. Judging by past experience, she should emerge sometime later in the afternoon, probably between six and 6.30 p.m., when the temperature had cooled. But within the hour, the nasal snort of impalas caught my attention. The Paradise female must have crept from her hiding place. I searched for the impalas with my binoculars, hoping that they would point me in the right direction. Perhaps they had spotted the leopard moving among the undergrowth, trying to turn adversity into a meal? A leopard is a past master at hiding from its enemies one moment and launching a fatal attack on prey the next.

Only a minute earlier the male impala had been reaching up to pluck a ripe acacia pod from the bush under which he had been ruminating. He probably never even saw the cat that killed him. The Paradise female had crept to within fifteen metres before bounding forward, catching the antelope unawares. A massive surge of adrenaline put springs in his heels as he leaped aside, but it was not enough. As he rose high into the air, two sets of claws clutched at his rump and held him tight. The impala fought for breath, the leopard's whiskered muzzle clamped over his windpipe, dagger-like canines piercing his throat. Reaching up with one forepaw, the Paradise female pulled the antelope down into the grass and pinned it to the ground, avoiding the impala's wildly thrashing legs. Gradually the frantic movements were

ABOVE: *Leopards prefer to hunt by stalking as close to their prey as possible, then rushing forwards at lightning speed, grabbing the animal's rump with their forepaws and biting into its throat to strangle it.*

OPPOSITE: *These male impalas are showing the typical 'alert' posture in response to a predator. Like other prey animals, impalas use alarm calls to let the predator know that it has been seen. The various prey species recognize each other's alarm calls and are alerted by them.*

stilled. After a minute or two the leopard released her stranglehold and looked warily around her. Then, satisfied that she had not been seen, she straddled her sixty-kilogramme meal and pulled it under the nearest bush.

Finally the day began to cool. A coqui francolin called, a lone voice in the stillness, proclaiming its territory. It was the last hour of the day. Like a wheezing concertina, a giant eagle owl let out a series of muffled hoots. Smaller birds heard its cry and chattered a scolding retort. Soon, when night came, the eagle owl would launch itself from its perch on silent wings to hunt for rodents. As the energy-sapping heat faded, the leopard's breathing slowed, the sharp colours of her coat grew indistinct, cloaking her for the night. The spotted hyaenas had already begun to stir in their dens, ready now to scour every part of the Paradise female's domain in search of food. Inevitably, the irresistible smell of a freshly butchered carcass would reach their nostrils. The crack and splinter of bone as the leopard fed would be music to their ears.

Half an hour before darkness fell, the Paradise female began to feed, gnawing at the impala's plump thighs. She sensed the danger. Hurrying now, she sheared through the thin belly skin, opening up the carcass so as to free the stomach and intestines and lighten her burden. She stood up and looked around, her abdomen bloated by her pregnancy, plate-like swellings visible above her teats. It was thirty metres to the base of the nearest tree and her prey was still too heavy to hoist safely above ground.

Next morning the sun lofted into a band of deep pink clouds. A lone hyaena crisscrossed the kill site, searching for scraps. Further up the ridge, towards Leopard Gorge, other members of its clan fought and cackled. All that remained of the Paradise female's kill was the impala's naked skull and vertebrae.

Cisticolas chattered in alarm as the leopard made her way back down the lugga. She moved carefully, wary of another confrontation with the hyaenas. Later she lay down to rest, her golden, spotted coat blending perfectly among the tawny grass-heads, catching the morning light. A hyaena came at the gallop, attracted from afar by the whoops and giggles. The Paradise female hissed out a warning as he moved purposefully towards her, but the hyaena barely faltered. The leopard's presence seemed singularly unimpressive to the burly predator, who easily outweighed his forty-kilogramme rival and had teeth to match hers. He circled downwind of the leopard, scouring the area, eager to track down the source of the meat smell. But he was too late.

The weather has teased unmercifully these past few weeks, the heat pounding like a throbbing headache against my skull. It is the time of the long rains, and the Masai sense disaster. Like the wildebeest, they are dependent on the vagaries of the seasons, their movements governed by the rains. Those living in the lower rainfall areas to the east of the reserve have endured drought for the last two months. People could be heard saying: 'Sell your surplus stock. Keep only the minimum that you need to get by. Do not risk losing everything.' But many Masai still cling to the belief that it is better to hold on to their cattle to the last, that God will answer their pleas for rain and their livestock will survive. Herdsmen trek with their cattle from more than a hundred kilometres away to graze within sight of the Mara. Tens of thousands of scrawny animals cluster along the reserve boundary. Each morning as I search for the Paradise female I find more people wandering across the plains. Before long the land seems empty of wild animals, filled instead with brindled cows.

If it does not rain soon hundreds of cattle will starve, and the wealth of the Masai will be reduced to ashes on the barren pastures. Councils of elders have already met to arbitrate on the pleas of their brothers for permission to graze their herds. The old men have seen it all before, bringing the wisdom garnered from years of drought and deluge, soothing the worried expressions of their younger relatives. 'Together we will survive,' they say. The elders nod in agreement. Tomorrow it could be them asking for grazing rights.

The *morani* quickly set to work creating circles of thorns, hacking down acacia bushes with their razor-sharp *simis*, the short swords traditionally used in hand-to-hand combat. The warriors' only concession to the chill night air is to construct spartan shelters within the thorn-bush enclosures, from where they can keep watch over their treasured animals. 'After all,' says one old Masai, 'they are *morani*, imbued with the strength and courage of youth. They do not fear the cold.'

Each day the clouds build into the sky, filling it within the promise of rain. The thunder rumbles like an angry lion. The herdsmen turn towards the gathering storm and breathe a sigh of relief. At last, during May, the skies open, unleashing the rain. Sheets of water quickly gather in the low-lying swells, forming bright pools fringed with green. Within minutes luggas are turned into torrents and I find myself trapped by the flood waters.

If the long rains have been slow in arriving, they are undoubtedly making up for it now.

Only the elephants continue to stroll across the plains: dark, lumbering shapes silhouetted against a grey sky, seemingly unconcerned by the water cascading from their backs. Thomson's gazelles huddle in parallel lines on the hillside, bunched one behind the other, trying to shield themselves against the elements. The rain lashes down harder than ever. Trees stand out like deathly shadows marooned in mist in the greening plain. Lions bow their bedraggled heads between their paws – a far cry from the king of beasts – while zebras stand as if chastised, their sturdy backs braced against the storm. Frustrated by the elements, visitors sit out their afternoon game drives, imprisoned by the steam of their own breath, peering into the gloom. And the Paradise leopard crouches, droplets of water clinging to her long white whiskers, staring from the darkness of a termite mound.

When at last the rain eases, the Tommies break rank and begin to feed. The lions shake themselves, sending plumes of spray into the air, suddenly awake to the chance of easy prey. Bat-eared foxes emerge from their damp underground dens, leaping high into the air to snap at the thousands of flying ants and winged termites fluttering above them. The young foxes race this way and that, sometimes pausing to lick the water from each other's fur. But the gazelles ignore their playful antics, displaying none of the wariness they exhibit in the presence of the predatory jackals, which are always alert for the chance to ambush a newborn fawn 'lying out' on the open plains.

For the moment the storm seems to have blown itself dry. Birds begin to call and gradually the thunder fades: the sky brightens. Within a kilometre of where I sit are eight lions, a mother cheetah accompanied by a year-old cub, and the Paradise female. For the moment all the cats are resting. But how different they are. The Gorge pride lie sprawled nonchalantly around the base of an acacia bush. Three young cubs wrestle among the giant paws of one of the lionesses, a huge creature. The cubs are relaxed and confident, basking in the powerful presence of their mother. A few hundred metres away, the cheetah female looks constantly about her, responding nervously to the slightest hint of danger, anxious for the safety of the sole survivor of her litter of five cubs. Meanwhile, the leopard, all fluid grace, skulks unseen between them, living out her secret life.

Even before the rain has stopped, the Paradise female has begun to work her way into position to take advantage of unwary prey. She has eaten little of any substance for the past two days, and now is an ideal time to hunt. She slinks from one patch of cover to another, stopping, looking around, keeping her head low. But when the rain begins afresh she abandons the hunt. With ears laid back and body pressed close to the ground, she rushes snake-like towards a tangle of thick bush, insinuates herself among the thorny branches and disappears from view.

By now, the Paradise female was nearing the end of her pregnancy. She looked uncomfortable, almost distressed, as if she did not quite know what was happening to her. I wondered if she was aware of the tiny cubs moving within her. She was restless, constantly sitting up to lick her teats or clean her beautiful white belly fur, which was plastered with mud where she had been padding around in the rain-soaked tracks. Everyone hoped that she would choose Leopard Gorge or Fig Tree Ridge to have her cubs. At both of these spots she could be viewed without being unduly disturbed. The timing was perfect. If she gave birth before the height of the tourist season, her cubs would be off to a good start.

ABOVE: *The saddlebill stork* (Ephippiorhynchus senegalensis) *is the largest of the Mara's storks and hunts for large insects, fish and frogs in ponds and marshes. Like all storks, the saddlebill is generally silent, though it uses bill-rattling as a greeting between pairs, and can also utter guttural sounds and hisses. The bird in the picture is a male; the female has yellow eyes and no wattle.*

RIGHT: *Egrets and herons are graceful birds, flying with their heads carried back on their shoulders with neck curved in a characteristic S-shape. Within the family Ardeidae, white birds are commonly called egrets, the rest herons. The great white egret* (Egretta alba) *is Africa's largest egret, a local resident throughout the continent, frequenting swamps, borders of lakes and marshes.*

OPPOSITE: *The grey heron* (Ardea cinerea) *is both a resident and a winter visitor throughout Africa, living on coasts and inland waters. It is most often seen hunched in the shallows, waiting patiently for an unwary frog or fish to venture nearby. Then it will creep forward stealthily to stab at its prey with its beak.*

But as the time drew closer the Paradise female remained on the low hill known as Emarti ya Faru, skulking around the rocks and acacia thickets a few kilometres to the south-east of Leopard Gorge. This was an ideal hunting ground for her, with fewer hyaenas and lions. Clustered on the green pastures clipped short by the cattle and wild herbivores, there were plenty of Thomson's gazelles and impalas – her most common prey.

One morning I found her in Leopard Gorge – or at least I thought it was her. I had been away from the Mara for the past two weeks and had returned expecting to find the Paradise female with cubs. As always, it was necessary to rediscover how to use my eyes again; to retune my senses to the way of life of the individual species I was looking for. When at last I caught sight of her, she was lying in the shadow of a deep cave with only her head protruding. Somehow, as she stared back at me with pale yellow eyes, she looked different. There was a wariness about her which I had never seen before. But then she had cause to be nervous, if she had cubs. I remembered how fierce Chui had been when Light and Dark were first discovered. All her protective instincts had been sharpened. The blood-curdling sight of her sprinting low and fast towards my vehicle remains with me to this day. The window was held wide open by a camera mount bolted to my door, but at the last moment she had pulled up, her yellow eyes blazing with fury. It had only been a mock charge. But it was a lesson I never forgot.

On finding the Paradise female, my first thought was to drive away so as not to draw attention to her. I knew that if I stayed in one spot for too long it would attract other vehicles, so I drove on down into the gorge. Only then, as I made my way past its overhanging rocks and fig trees, did I suddenly realize that this was not the Paradise female at all. It was the male cub whose mother shared part of the same home range.

The young male had grown considerably since I last saw him at the Cub Caves on Fig Tree Ridge. At one year of age he was more than two-thirds the size of the Paradise female, though his coat was much darker than hers and his face was longer. He seemed nervous as I approached, but held his ground as I edged forward for a closer look. I was delighted to see him again and to know that he had survived. But his presence in the gorge meant that the Paradise female was almost certainly some distance away.

I decided to drive to Musiara Marsh, intending to wait until later in the afternoon before returning to Leopard Gorge. But just as I was preparing some lunch in the back of my car, a vehicle drew alongside. It was Samuel Langat, an old friend among the drivers at Governor's Camp. Langat told me that earlier in the day he had tried to signal to me, flashing his lights to attract my attention. He had found the Paradise female with an impala kill stored in a tree on Emarti ya Faru. But the excitement of discovering the leopard with a kill was tempered by Langat's assertion that she had lost her cubs. Apparently she had been seen the previous day emerging from a deep thicket, looking very agitated. Perhaps her cubs had been stillborn, or she had chosen her den unwisely: it is not uncommon for domestic cats and dogs to fail to raise their first litter due to inexperience. Lions or hyaenas – even baboons – would certainly penalize the slightest error of judgement on the part of a mother leopard who failed to conceal her newborn cubs carefully.

A few months later, during the dry season, an event took place that was to have a marked effect on the lives of the Paradise female and the young male leopard from Fig Tree Ridge, invoking the age-old conflict between man and the predators.

The two boys moved lightly across the plain, driving their fathers' cattle to water along the Ngorbop lugga. The pace of the cattle quickened a stride as the tantalizing smell of water reached their nostrils. It was a journey they had made many times before; but the intermittent water-course had withered these past few weeks in the unrelenting heat. One of the boys wandered further downstream, searching for another pool to quench the thirst of the herds.

Leopard females usually give birth to two or three cubs, but in areas where there are many competing predators, they do well to raise even one of their litter to adulthood. This picture of Chui and Dark on Fig Tree Ridge was taken in 1983: the leopard family disappeared shortly afterwards and there is no way of knowing whether Light and Dark survived to become independent of their mother.

Leopards can go without drinking for days at a time – in areas such as the Kalahari, they avoid heat stress by remaining primarily nocturnal, thereby conserving water. Although it is rarely necessary in the Mara, they can survive without drinking at all, obtaining sufficient liquid from the blood and body fluids of their prey.

ABOVE AND OPPOSITE: *Cattle are still an integral part of the Masai's life, but men like Ben Kipeno, co-ordinator of the Friends of Conservation's Community Conservation Programme, are also concerned with the wild animals. The future of the Mara region lies in the hands of educated Masai like Ben, who can communicate the role of wildlife-based tourism in conservation.*

He moved carefully, pausing every so often to peer below the clumps of bush that crowded the lugga, taking care not to disturb a resting buffalo or pride of lions. Suddenly he felt the tight curls of hair tingling down the back of his neck. He stopped, acknowledging an uneasiness that was second nature to a child raised among wild creatures. Peering into the dark recesses of the bush, he concentrated his vision to try and distinguish what it was that made him feel so uneasy. He knew, instinctively, that something was terribly wrong.

The old male leopard had heard the warning sounds of Masai cowbells and high-pitched whistles. Having grown accustomed to living among the Masai for the past ten years, he chose to ignore the cattle as they drifted towards the lugga. They were still too far away to cause him to abandon one of his favourite resting places. The leopard gambled on his ability to remain undetected, never moving unless he felt he had been seen. By the time he opened his eyes again and looked up, the herdboy was staring right at him. It was too late to flee. He crouched, every muscle of his body ready to attack.

Suddenly the boy realized what he was looking at. Fear gnawed at his stomach as his brain struggled to take in the horror of what he now saw. The dappled pattern of light and shade between the fork of the fallen tree was the subtle camouflage of the leopard's coat. The boy stifled the almost irresistible urge to run, calling out softly to his friend. As the second boy

Masai men traditionally wear a red robe or shuka. *This photograph was taken at a ceremony to 'name' a new generation, prior to the boys' circumcision. After they are circumcised, the boys become warriors or* morani. *But times have changed – there are no longer wars to be fought against other tribes, and cattle-rustling is a serious crime.*

came into view, the leopard streaked low across the ground. The blinding speed of his charge gave the herdboy no time to flee or defend himself. It was like a bolt of lightning – sixty kilogrammes of pent-up fury exploding from under the fallen tree. At the last moment the leopard rose and reached for the boy, raking him across his stomach and legs; then just as quickly he hurried back to the bushes. In a state of shock, the boy managed to crawl as far as the herd of cattle, huddling among their legs for safety. The second boy, meanwhile, had fled for his life, weaving his way through the cattle until he was safely out of sight.

Racing to the nearest manyatta, the boy blurted out his story. Four men came running, armed with spears and the courage that had been tested in the face of the terrifying charge of a full-grown male lion. Masai blood had been spilled; revenge had to be taken. This was a question of honour, a chance to prove themselves. The men could have chosen no more determined an adversary than a cornered leopard. He was everywhere, so fast and agile that he had clawed and bitten two of the Masai before they even realized they had been mauled. One of the men threw his spear and missed. The next moment his lip had been sliced open, a tooth torn out. He lashed out with his *simi*, hacking at the cat's thick muscular neck as it bit and clawed him. Another man drove a spear into the leopard's abdomen; but it was not enough, and the injured cat retreated to cover.

By this time more than a dozen men had gathered in a fever of excitement. The predator must be vanquished at all costs. The men formed a broad front, trying to drive the wounded leopard into view, but he was too cunning. He snaked back and forth through the dense cover, never providing a clear target. Guided only by the animal's furious coughs and explosive grunts, some of the men threw their spears in his general direction. The leopard charged again, and this time the men fled; for once their redoubtable courage had deserted them. Despite his injuries, the male quickly gained on the last man, clawing at him, but in the process was speared again. Mortally wounded this time, with at least one spear hanging from his body, the leopard was left to die and the Masai withdrew to count the cost.

The injured men were taken by car to Mara Rianta, where a small dispensary had been built some years earlier with the proceeds of tourist revenue. Wounds inflicted by predators are notorious for becoming infected, so it was essential that the men's injuries be cleaned and that they be given powerful antibiotics to prevent blood poisoning. All eventually recovered. Next morning the rangers came to look for the leopard, but could find no trace of the body. In all probability the hyaenas had disposed of him.

The male was well known to the drivers. It was he who had mated with the Paradise female in February. His territory included both her home range and that of the mother of the young male from Fig Tree Ridge. His death meant that other males would quickly compete to establish their own claims to his territory, and the right to sire the Paradise female's next litter.

—8—

Nomads of the Plains

How can one convey the majesty of the Serengeti? . . . The magic of the unraped American prairie here blends with the other magic of the animals as they existed before man. There is a lightening of the spirit, a sense of atonement, of being able to make amends at last for the endless cycle of our vanity and greed to which they have been subjected. Nature's world of beauty and justice without cruelty or compassion.

THE SUNDAY TIMES
CYRIL CONNOLLY

THUMP, THUMP, THUMP. The familiar sound of large animals fleeing greets my ears as I emerge on top of the escarpment. Probably waterbuck, I think to myself, exhausted by the final push to reach the rocky summit. But it is not: four tawny shapes run for cover, swallowed up by the long grass. The young lions rumble among themselves – as much to each other as to warn me of their presence. I watch them go, thrilled to see lions at such close quarters. They are sleek and golden – young males of two to three years of age, with scruffy blond manes. I walk to the place where the lions had sprawled on their sides, heads curled. Four man-sized depressions flatten the ground – lion-shapes moulded in grass. I kneel down and press my face to the grass, drawing in the strong musky odour of the cats. Three of the lions had lain close together, the fourth a few metres away. I retrace my steps. Just twenty paces.

There is no doubt in my mind as to the identity of these lions. One more question has been answered. They are the four young Kichwa Tembo males, who disappeared from the pride territory months ago. Somehow they have survived among the Masai and their cattle, avoiding the poachers, enduring the long wet season since the Old One perished. Their mothers, having suffered the loss of Black Beard and Blond Mane, had eventually returned to the Kichwa Tembo pride territory and mated with the Bila Shaka pride males. Soon there

OPPOSITE AND OVERLEAF: *Swimming the Mara River is hazardous for the wildebeest, but in a dry year it is extremely shallow and the herds can walk through knee-high water with little or no loss of life. Large numbers cross in a matter of minutes, though the limited perceptions of the wildebeest still make them vulnerable to crocodiles.*

Two one-and-a-half-year-old males from the Kichwa Tembo pride. At this age they are almost independent enough for their mothers to come into oestrus again and produce a new litter of cubs, but they are still inexperienced hunters and it will be many months before they are able to survive on their own.

would again be cubs for me to watch along the Sabaringo lugga. But there was no place in the pride for the four young males. They were destined to wander as nomads for the next year or so. If they were lucky, they would live long enough to take over a territory of their own.

The silence is broken by a soft moan of greeting. It is the lions. Perhaps they have reunited among the thickets behind me, reassuring each other after the shock of my sudden appearance. A scaly francolin scolds as it streaks through the grass. I sit for a while longer looking out across the plains. Thirty kilometres to the south lies the great land of Serengeti. The mere thought of that wild place fills me with excitement. It is as if I can already see the wildebeest and zebras, can hear their flying hooves beating a tattoo across the plains, as two million animals wend their way northwards. Will it be a 'good' migration, I wonder – will they mass together, arriving in their tens of thousands, or will it be a mere trickle compared with drier years? Trying to predict the extent and timing of the migrants' annual journey through the 25,000 square kilometre Serengeti–Mara ecosystem is pointless. Each year is different, reflecting the vagaries of the seasons. All I can do is to wait and see.

Six months into the year and the rains are meant to be over. But nobody believes it for a moment. These past three years in the northern Mara are remembered for the amount of rain that has fallen in both wet and dry seasons. The wild fields of red oat grass have remained long and rank, waiting for the migratory herds that never came. Fires mask the sky with tall plumes of acrid smoke that rise like a giant atomic mushroom, gradually fading into a grey blanket that hangs over the plains. A pall pervades the atmosphere, fortified by fires set in the great papyrus marshlands of the Sudd in the Sudan, and in the savannahs of Zaire. Throughout central Africa if it the time of the great burning, the laying to waste of the rank grass.

By the end of June the advance parties of zebras and wildebeest have swept in a great wave across the Sand River, the intermittent water-course which marks the boundary between the Serengeti and the Mara; perhaps this year they will stay for a while. The bull wildebeest are as frenetic as I have ever seen them, reaching a peak of activity during the annual rut. Territory holders noisily advertise their presence, advancing determinedly on anything that they perceive as a threat. One wild-eyed animal lunges to within metres of my vehicle, scything his sharply pointed horns through the air before racing off in the realization that the green truck is not, after all, a member of his own kind.

To the casual observer the rut appears to be a scene of total chaos. Bulls rush with dark eyes bulging, patrolling their makeshift territories as they attempt to mate with as many receptive females as possible. The cows – the majority of which are accompanied by four-month-old calves – cluster in groups up to thirty-strong on the males' territories, waiting for a chance to move on. But there is no escape from the attentions of the bulls. The synchronized timing of the rut ensures that 400,000 calves will be born within a few weeks of each other during February and March, when the rain-swept plains of the southern Serengeti are at their most nutritious. The species of short grasses found there are sown on ashy soils derived from the ancient volcanoes of the Ngorongoro Crater Highlands. They are rich in calcium and sodium, vital nutrients for lactating mothers and newborn calves. This is the reason the wildebeest leave the Mara at the end of the long dry season in October and November and hurry back to their ancestral calving grounds, 200 kilometres to the south of where I stood.

The migratory wildebeest follow two distinct routes as they abandon the Serengeti Plains at the beginning of the dry season, forced to search elsewhere for grass and water. Many wildebeest take the traditional westwards route through the whistling thorn country of the Western Corridor, running the gauntlet of the meat poachers and lion prides. By early August they have exhausted the best of the grazing in the corridor and are obliged to turn north to brave the icy waters of the Mara River. The survivors of the river crossings are rewarded with hundreds of square kilometres of rich grazing in the Mara Triangle.

But as the migratory population has expanded to current levels, some herds have begun to follow a different path. The animals that I now watched from a rise above Keekorok Lodge, in the southern part of the reserve, had travelled directly north after leaving the desolate Serengeti Plains at the end of May, avoiding the Mara River to the west.

Though the first wildebeest often arrive in the Mara by late June or early July, it is not until some weeks later that they are to be seen in the most northerly part of their range. Then, long lines of animals plough a broad furrow through the waist-high stands of red oat grass around Kichwa Tembo, while others stream over Rhino Ridge towards Musiara Marsh, where the Bila Shaka pride are lying in wait.

By August a familiar question is on the lips of every visitor to the northern Mara. 'Where is this migration?' Some might

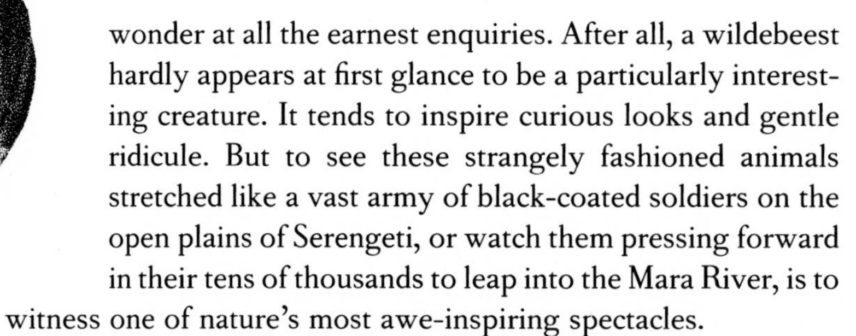

wonder at all the earnest enquiries. After all, a wildebeest hardly appears at first glance to be a particularly interesting creature. It tends to inspire curious looks and gentle ridicule. But to see these strangely fashioned animals stretched like a vast army of black-coated soldiers on the open plains of Serengeti, or watch them pressing forward in their tens of thousands to leap into the Mara River, is to witness one of nature's most awe-inspiring spectacles.

To everyone's disappointment, the pattern of the previous two years repeated itself yet again. The Serengeti wildebeest were slow to arrive in the northern Mara. Heavy rains in late June had allowed the herds to disperse widely throughout the Western Corridor, and those animals which had already begun to move directly north only tarried for a week or two in the Mara before heading south again. They preferred to congregate where the grass was short and richest in protein, rather than to plunge needlessly into the longer grass ahead of them. Fortunately for the Bila Shaka pride and their nine large cubs, thousands of zebras had surged into the Musiara area from the northern Serengeti, mingling with the herds of migratory wildebeest and zebras from Kenya's Loita Plains. But although there was plenty of food for the pride there was danger, too.

Accompanying the wildebeest and zebra were the inevitable nomadic lions – bands of males that each year arrived in the Mara with the wandering herds. Compounding this was the fact that over the last few months the Bila Shaka pride males had responded to the loss of Black Beard and Blond Mane by increasingly making incursions across the river in the Kichwa Tembo territory. Having gained access to both groups of females, the Bila Shaka males chose to spend most of their time on the Kichwa Tembo Plains, abandoning Grey Coat and her relatives for long periods at a time, creating an uneasy vacuum in the Musiara area that would not go unchallenged.

The benefits of living in a group, which distinguishes lions from other cats, are not simply that it is easier to gain a meal or that together they can overpower prey the size of an adult buffalo; a solitary lioness has just as much chance of feeding herself. By living in a group, lions have greater success in defending their kills against the more numerous hyaenas, which are always on the look-out for an easy meal; if one must share, then it is preferable to share with a relative. But perhaps most important of all, being a member of a pride means that individual lionesses are better able to safeguard their genetic interests, helping to protect their own cubs – and those of close relatives – against male intruders.

The sight of vultures plummeting to earth caused the Bila Shaka lionesses to look up. It was not just the possibility of scavenging a meal that made them get to their feet and move uneasily towards the place where the vultures were crowding into the scattered trees out on the plain. The four lionesses sensed danger in their midst. A kill had been made in the heart of their territory. As they drew closer, two blond-maned lions rose out of the grass in front of them. The sudden emergence of the alien males caused Grey Coat to stop and glare. Lowering her head and flattening her ears, she hissed out a warning to them to move away. Brown and the other lionesses quickly moved up to join their older relative, galvanized by the urge to protect their territory; never more so than now, when their own pride males had abandoned them.

The two young Serengeti males were about three and a half years old, considerably larger than the adult lionesses they now faced. Since they had left the pride of their birth, the young males had been forced to look after themselves, competing at every turn with other

A white-backed vulture approaches a kill. It can be distinguished from the closely related Ruppell's griffon vulture by its black bill – the Ruppell's has an ivory bill. There are six species of vulture in the Serengeti-Mara, a total of 40,000 individuals, and each year they feast on more than twelve million kilogrammes of meat!

In the Serengeti-Mara the white-backed and Ruppell's griffon vultures make up eighty per cent of the birds round a carcass. They follow the movements of the migratory herds, while the territorial lappet-faced and white-headed vultures are confined to a more limited area. Large numbers of Ruppell's nest in the Gol Mountains to the east of Serengeti and may travel a round trip of more than one hundred kilometres in search of food.

ABOVE: *Lions use a variety of vocalizations and facial expressions to communicate with one another, thereby minimizing the need for violence and the risk of injury. Lionesses will try to defend their cubs when challenged by nomadic males and are sometimes seriously injured in such encounters.*

OPPOSITE: *By the time a male lion is three years old, he will normally have been evicted from his natal pride (with any brothers or male cousins of a similar age) to lead the life of a nomad. Now he must not only hunt for himself, but must try and avoid conflict with other lions, at least until he and his companions are old enough to challenge for a territory of their own.*

predators. But now that they were almost fully grown, they were rapidly gaining confidence in their own power. Already they had begun to mark and scrape, announcing their presence in a new area. Soon they would challenge for a territory.

The males had followed the wildebeest herds north from the Serengeti. By attaching themselves to such an abundant food source they had lived well, either by killing for themselves or by scavenging from the hyaenas. As they moved with the wild herds, the two lions trod gingerly through the minefield of pride territories. Cautiously they had probed, testing the strengths and weaknesses of the various prides, hurrying through the heavily defended territories around Keekorok and Talek, where prides number up to forty individuals apiece with three or more males occupying territories of fifty to a hundred square kilometres. As yet the young males were unmarked by serious battle; and here at Musiara Marsh they sensed that it was safe to stay for a while. They had watched and listened from the

high ground above the marsh. There were times, as they stared into the distance, when they had seen the zebras scatter and run as the Bila Shaka lionesses ambushed the herds. They had found no sign of the two big males, though they had heard their roars echoing back across the river, hinting at ownership. But for the moment, the marsh and its surrounds were wide open to intrusion.

At first one of the young males held his ground, standing and staring back at Grey Coat; but once she broke into a run he turned and loped away after his brother. She pursued him with a vengeance, her younger sister, Brown, hurtling through the long grass to join her. The speed and aggression of the charge were electrifying. Brown thundered past Grey Coat, not stopping until she had closed right in behind the male, grunting with anger. But she resisted the temptation to rake her claws across his rump. He was bigger and stronger than she was and, one on one, might easily injure her if forced to turn and defend himself.

The threat of violence and the collective strength of the four Bila Shaka lionesses were enough to drive the males towards the murram pits bordering the Governor's Camp airstrip. Eventually the lionesses led their cubs away; but as soon as they turned their backs and began to retreat, the two males rose to their feet and followed. Seeing the males advancing once again, some of the younger lions turned to meet their challenge, emboldened by the presence of the adult lionesses. The more aggressive of the two males paused, raking the ground with his hind claws, tearing up the soil and urinating as he glowered defiantly at a young lioness moving towards him. The Serengeti males did not fear her. In fact the sight of her stalking towards them only incited them further. But before they could attack her, Brown and Grey Coat came charging to the rescue, grunting and snarling as they rallied to the defence of their young. The males did not stay to put their challenge to the test. This time they were outnumbered; but they would be back.

The Bila Shaka lionesses retired to the patch of forest near the southern end of the marsh, where they had rested earlier that morning. During their absence an old buffalo had moved from the place where he had been feeding and walked out into the open. He lay there in the long grass for more than two hours. Now, as the lions slept, he was on the move again. Long scratches marked his dusty hide, as if he had been raked with claws. Had he already survived a half-hearted attempt on his life by the young Bila Shaka lions?

Wearily the old bull buffalo trudged along the edge of the forest. The thick, rounded chunks of his thigh muscles had wasted away, exposing the bony contours of his pelvis. His mud-caked hide hung loosely from the crest of his spine and the gauntness of his body somehow made his widespread horns look even more impressive. They were scored and chipped at the boss, sweeping down and then up to the great curved points which glinted in the morning light. In the days when hunting was permitted in Kenya, such a trophy would have drawn admiring gasps from any professional hunter.

The bull was in his fifteenth year and no longer submerged himself within the protective presence of the herd. It was too much effort to try and keep up with the healthier animals; and besides, he had developed a singular routine that suited his advanced years. Each morning he made his way along the forest edge, until he reached a small patch of marsh hidden within a dense clump of trees; the same place that Khali, the Marsh lioness, had frequented before she disappeared. Lowering himself over the steep edge, he would slide forward until his front feet reached the water, easing himself into the soft mud.

Four hundred metres away, the Bila Shaka pride lay sleeping. They had failed to notice the old buffalo as he entered the marsh and began feeding on the lush vegetation sprouting from the bank surrounding the pool. Dense clumps of sedge concealed him from view. A black crake dived in and out between his feet, snapping up insects and frogs with its vivid yellow

A young baboon clings to its mother's belly fur as she searches for food and tries to avoid predators. When it is three months old, it will start to ride 'jockey-style' on her back, every so often climbing down to feed on the vegetation or scamper around. At night the troop retreats to trees or rocky cliffs as a refuge.

bill. Nearby, a rufous-bellied egret stood watching, motionless. The bull fed quickly, his long grey tongue wrapping itself round mouthfuls of soup-like water-weed. When a baboon barked out a warning he did not even bother to look up. The old bull knew he was safe as long as he stayed close to water.

The lions flowed like thick brown treacle from one patch of shade to another. At first they were restless, pausing to rake their claws on the rough bark of the trees. One of the lionesses hauled herself in a most ungainly fashion up into the lower reaches of a sapling, hugging the trunk and then slithering down again. What a contrast to the fluid grace of the Paradise leopard. A young male crouched over an abandoned skull from an old zebra kill, squabbling with whoever tried to join him. One of his sisters stalked another, leaping forward and bowling her sibling on to her back. The scene played itself out in a series of disjointed moments. A flash of lilac and blue streaked overhead as a roller dive-bombed a woodland kingfisher, grating out its strident call. Later, as the lions moved towards water, the baboons

ABOVE AND OPPOSITE: *These zebras and wildebeest have just crossed the Mara River, using the narrow hippo trails as exit points. Although they have safely negotiated the most dangerous part of their migratory journey, they may still fall victim to the ambush of a lion or a poacher's snare.*

barked and shrieked again, clambering into the tops of the tall African greenhearts, shaking down a hail of plum-sized fruits, which fell to ground with a resounding thud and were gathered up by other members of the troop.

As shadows shortened towards midday, the Bila Shaka pride clustered beneath a single tree, sprawling with legs and paws draped comfortably across each other's midriff or backside. Some lay with their snow-white bellies turned skywards, reflecting the sun. Brown rolled and squirmed, working a patch of skin that itched, then turned on to her side again, reaching out in half sleep to make contact with another pride member. Only the flick of a black-tipped tail or the twitch of a large, round ear indicated that the lions were there. A bull giraffe glided by on stilt-legs, pausing to stare down on the mound of lions, then drifted on to the place where the old buffalo lay up to his elbows in soothing mud. By now the lions were only 150 metres away, but still the old bull was unaware.

As the heat of the sun began to fade, the buffalo moved to the edge of the forest, where he lay among the long grass, rhythmically regrinding his food with the stumps of his worn-down teeth. But when vehicles approached him during the afternoon game drive, he rose, unsteady on his feet, and headed back to the tranquillity of the marsh. Once more he felt secure in the water, staring out through clouded eyes as he munched at the soft vegetation. He and the lions were playing a deadly game of hide and seek of which neither was aware.

The Mara is a vital dry season refuge for the wildebeest. Here they are assured of plentiful grazing until their instincts tell them it is time to return to their ancestral calving grounds on the Serengeti Plains, two hundred kilometres to the south. There hundreds of thousands of females will give birth within a few weeks of each other, and the whole cycle will begin again.

The old and the sick are ideal targets for any predator. The nine cubs would relish the chance to toy with the bull, leaping on to his back and then dragging him to his knees.

All day long the lions drifted in and out of sleep. Every so often one or other of them would sit up and look around, briefly inspecting the herds of zebras and waterbuck gathered where the grass was still green and lush. As the air grew cooler, one of the young lionesses crept out to the edge of the plain. The pride always took particular notice whenever a party of zebras drifted towards them through the long grass. The zebra herds had been a welcome addition these past two months, fortifying them while the wildebeest were in such short supply. But, as so often happened, inexperience let the lioness down: she charged towards the zebras before she was close enough to have any real chance of success.

At nearly two years of age, there were marked differences between the nine sub-adult males and females in the Bila Shaka pride. This reflected the differing roles that males and females play as adults within lion society. The five young lionesses were noticeably more alert to the possibility of hunting for themselves. There was a sharpness, a sense of purpose that was missing from their larger, slower brothers. When the pride moved, the young females often kept the momentum going, particularly when the enthusiasm of the adult lionesses flagged. In contrast, the young males were more aggressive in their play and used their greater size and strength to good advantage at a kill.

At five o'clock, the old bull hauled himself out of the water for the last time that day and headed back along the edge of the forest. He stood for a moment, his large wet nose held into the breeze; but he was too far away to pick up the lions' scent. At first the pride did not see him plodding towards them, but when they did, there was no doubting their intentions. One by one they looked in his direction, ears cocked and mouths tense. On he came, veering suicidally towards the thicket where the lions were waiting.

The old bull was almost on top of them before he realized his error. Too late to run now. He was trapped within a circle of lions. He grunted and clicked, charging blindly towards his adversaries. When they closed in he stood his ground, defying them to come closer, hooking and thrashing the bushes with wild, swiping blows of his heavy horns like a butcher sharpening his knife. The speed and power of those terrible horns – swung from ground to shoulder – could easily toss a lion high into the air. The young lions needed no further proof of the buffalo's legendary willingness to be joined in battle; if they had thought that this was going to be an easy kill, they soon realized they were wrong. Try as they might, they could not get in close behind the bull to launch themselves at his vulnerable rear. The youngsters wound around the older members of the pride, rubbing up against Brown and Grey Coat, soliciting their help, hungry for the kill, yet unsure what to do next. The lionesses responded with yawns of indifference and snarls of irritation; they had already decided that this was not the time. They were not hungry enough to risk death or injury on the points of those horns.

Then, just as the whole pride was losing interest in the confrontation, the old bull lay down. As one the lions rose to their feet. Had the moment they had been waiting for arrived? But even as Grey Coat began to circle, he stood up again, nose thrust into the air, sweeping his head from side to side, saliva welling over the sides of his slack jaws, until the pride backed down.

To the north, above the spring, a spiral of dust swirled into the sky, pinpointing the stream of wildebeest galloping in single file through the long grass. It was a sight to make any lion leap to its feet: they were headed straight for the Bila Shaka pride. Brown was the first to see

them coming. At once she turned and bounded through the grass, desperate to intercept them. For a moment the old buffalo just stood, confused, as the circle of lions evaporated from in front of him. Then, sensing that freedom was at hand, he trundled off through the bushes, heading for the safety of the river.

That night one of the lionesses killed a zebra within a hundred metres of where the pride had surrounded the buffalo; but by morning virtually nothing was left of it. One of the pride males had recrossed the river during the early evening and sated himself at the carcass. I looked for the old buffalo, but could find no trace of him. Perhaps he had decided to stay near the river or was skulking in the thick forest. But next day he was back, though this time he wisely kept close to two other bulls. He no doubt felt a sense of security in the presence of his hefty companions, who would help protect him if the lions decided to challenge him again.

Shortly after 9 a.m. the old bull got to his feet and began to head for the forest. His hunger drove him onwards, despite the warning barks and squeals of the ever-vigilant baboons. The troop had roosted in the tall trees during the night and were now trapped there by the proximity of the lions. But the bull could wait no longer. The urge to feed on the soft vegetation in the marsh was irresistible. He paused at the edge of the forest where the pride rested, waiting as one of the other bulls trudged alongside him. Just then the baboons began to bark with a renewed sense of urgency. The pride male was on his feet, wandering in full view towards the place where the lionesses and cubs watched. The big lion ignored everything, panting rapidly, his stomach bloated with meat. He flopped on to his side, forcing a great *umph* of air from his lungs, paying no attention to the baboons or the buffalo.

With the previous encounter still fresh in his mind, the old bull turned and hurried away a pace or two. Wearily he lay down facing the pride, who by now barely seemed to notice him. Meanwhile the baboons were restless. They were eager to be out feeding on the plains. They could not afford to lie around for most of the day and night as the predators could. The baboons needed to scour the plains for grass and insects, pilfer the forested areas for fruits and other delicacies, even on occasion surprising a gazelle fawn or impala calf. But every time one of the troop began to descend from a Diospyros tree, one or other of the young lions would see it and crouch, ready to charge. In the distance, the baboons could hear a party of ground hornbills uttering their resonant, booming call as the giant birds strutted across the marsh. Zebras and waterbucks gathered nervously along the edge of the forest, looking for the safest route to water.

At last the lions succumbed to the irresistible urge to sleep. One by one the baboons climbed into the lower reaches of the foliage, then shinned down the vertical trunks. The bolder among them ran from one tree to another, like children playing hide and seek, always making certain they kept a firm grip on the surest route to safety. Sensing that the moment was right, the baboons raced away, tiny black babies clinging to their mothers' bellies. The older youngsters rode on their mothers' backs, poised in the manner of champion jockeys hurrying past the finishing post. Some of the juveniles – noisy, boisterous individuals – tarried awhile, watching the rest of the troop gallop away to the marsh where the old buffalo liked to feed. Finally, the last of the troop dropped like ripe fruit from the trees in the certainty that they could outdistance the lions.

As the dawn welcomed another day, vultures gathered on the bare ribs of a topi carcass, unopposed by the pride, who languished in the long grass beneath a cloud-filled sky in which the sun was merely a distant glow. To the east, just beyond the Miti Mbili lugga, a fire still

The Bila Shaka lioness Grey Coat brings down a zebra. Zebras are commonly taken by lions in the Mara, particularly when the wildebeest are absent. But zebras have a potentially lethal kick and a lioness must be careful to avoid the flailing hooves as she grabs the zebra by the rump. A well-placed kick could break her jaw and leave her to starve to death.

raged, dancing across the plain like a great black ship, quickly gathering speed as the wind intensified. Soon the marsh was cloaked in a blanket of smoke. Elephants wandered through the long grass and into the cooling water of the reed-beds. Beyond them zebras and topi gathered to feed on the green shoots springing from the blackened earth. The light was flat, shapes dulled and lifeless, the heat pressing down across the land, trapped between sky and plains. African sand martins flitted overhead, reaping a rich harvest from the hordes of insects trying to escape the flames.

The lions retreated to a dried-up reed-bed near the main road to Governor's Camp. A large herd of buffaloes brooded to the south. Gradually the herd grazed towards the place where the lions were resting. Running this way and that through the tight ranks of buffaloes came a wildebeest calf. Born on the Serengeti plains in February, the calf was now six months old, the bases of his pointed horns just beginning to curve. Though the calf had evaded the predators and meat poachers on the journey north through the woodlands of the Western Corridor, he had recently become separated from his mother. Without her he was confused and vulnerable; calves normally continue to follow their mothers until they are a year old. His lean brown form stood out among the solid black mass of the herd. As he galloped about, weaving between the thicket of hooves and horns, one or other of the buffaloes would shoo him on his way, irritated by his skittish presence.

A zebra kill is large enough to provide the whole pride with food. It is often feast or famine for the predators, so they have the capacity to gorge themselves. A lion's daily requirement is about eight to ten kilogrammes of meat, but a male can consume up to forty kilogrammes on an empty stomach.

All this the lions watched from the mound, their golden yellow eyes following the calf's every move, anticipating an easy kill. It was as if the buffaloes did not exist. But just as it seemed certain that the calf would blunder headlong into the jaws of the waiting pride, the wind changed and the buffaloes caught their scent. The herd turned in panic, storming away from the reed-bed in a cloud of dust with the young wildebeest somewhere in their midst, leaving the lions squinting and bemused.

That afternoon the stream of vehicles visiting the Bila Shaka pride barely paused long enough to take pictures. Something was happening elsewhere, and it did not take much imagination to work out what it was. As one driver proferred in Swahili: 'There is a "wedding" at Leopard Gorge.' In other words, leopards mating.

The driver told me that during the last few days they had tracked the Paradise female and her consort as they moved from the Ngorbop lugga (to the east of Emarti ya Faru) as far as Leopard Gorge, two kilometres further north. He said that there was some confusion as to the male's identity; territorial males rarely offer a good view of themselves, shunning the sight and sound of vehicles. Apparently he was not a particularly big leopard, and certainly not as shy as the old male that the Masai had speared. On hearing this news, I immediately abandoned my vigil with the Bila Shaka pride and hurried towards Leopard Gorge.

—9—

Sunlight and Shadows

The highest wisdom has but one science – the science of the whole – the science explaining the whole creation and man's place in it.

War and Peace
Leo Tolstoy

When I arrived at Leopard Gorge, I found the Paradise female carefully grooming herself in the long grass. She licked her genital area repeatedly, then rolled on to her back as if scratching an itch – behaviour typical of a female in oestrus. On a cool, overcast day such as this she might well have taken the opportunity to hunt; but her hunger was stifled by the overriding urge to mate. The male, meanwhile, rested out of sight behind a curtain of dense foliage. It was a familiar hiding place, where I had found the young male leopard from Fig Tree Ridge on a number of occasions during the last six months. But surely this could not be him? He was far too young to mate and would be severely chastised by any adult male that picked up the scent of the Paradise female while she was in oestrus.

Again and again during the next few hours the Paradise female leaped up to the cave where the male lay. She slunk back and forth in front of him, wafting her scent in his face, crouching before him, inviting him to mount her. But the male seemed strangely reluctant, as if unsure quite what to do. He greeted her advances with explosive coughs and deep rumbling growls; but the tell-tale, stuttering outburst of sound when a male leopard climaxes was missing.

As the temperature began to drop, smoke sank into the belly of the gorge, blanketing the scene with an eerie light. Finally, at 6.15 p.m., the last of the vehicles hurried away, their drivers anxious to be back at camp before dark. Now the silence was total, broken only by the strangely mournful *tok, tok, tok* of Gabon nightjars.

OPPOSITE: *Having eaten, a leopard fastidiously cleans blood and meat from its face and paws using its rough tongue. One often sees cats with ticks clinging to their fur, which they try and remove by biting or scratching. The long whiskers which are clearly visible in this picture enable a cat to feel its way in the dark.*

Leopard Gorge is ideal leopard country, providing places to lie up during the day and relatively safe hideaways for cubs. Over the years a leopard becomes intimately acquainted with every aspect of its home range, which for a female may be ten to twenty square kilometres. A male leopard defends a territory which may encompass all or part of several females' overlapping home ranges.

Next morning I searched the area surrounding the gorge, but could find no sign of the cats. Knowing how easy it is to drive right past a leopard in thick cover or dense shade, I continued the search. Sure enough, as I drove back towards the gorge, I could see the leopards lying together on a broad slab of rock, warming themselves in the morning sun. Now was my chance to take a proper look at the Paradise female's suitor; I hoped that I was about to feast my eyes on a full-grown male. But he was barely as large as she was.

The coats of the two leopards were quite different in colour. Hers was washed with a rich golden brown, while he was darker, the coal-black spots and pale cream of his coat giving him an almost olive sheen. There was only one leopard in this area that fitted that description. It *was* the young male born on Fig Tree Ridge fifteen months earlier. During the next year his forehead would broaden and he would acquire the thick-set, muscular physique of an adult male. In the meantime he still had a lot of growing to do.

Until recently the young male had spent most of his time in the company of his mother, following her as she searched for prey, learning his way around her home range, part of which overlapped that of the Paradise female. But during the last few months he had spent more and more time alone, hunting small animals such as hares and hyrax, guinea fowl and grass rats. Then one day, when he was almost fourteen months old, he managed to kill an adult female impala, though he quickly lost it to a spotted hyaena as he struggled to haul the carcass towards a nearby tree. His dependence on his mother was at an end. Now he was embarking on one of the most dangerous periods of his life. The road to independence was strewn with casualties; many sub-adult leopards do not survive. No longer could the young male rely on his mother for food or protection; she would soon become pregnant with her next litter and would not welcome his company or desire for contact. He was on his own, and would need all the experience gained these past fifteen months if he was to survive. Now that the old territorial male had been killed by the Masai, another adult leopard would soon emerge, ready to fight to breed with females in the area. Baboons would continue to harass him whenever their paths crossed, and the Gorge pride would kill him if they could. And there were always the Masai to be avoided.

The young male was no stranger to confrontations with man. A few months earlier a Masai honey hunter had wandered along the base of Fig Tree Ridge, searching for wild bees. He visited a number of likely trees, checking the undersides of branches that he had hacked open in the past to encourage the bees to swarm. But on this particular afternoon, he had found nothing. The honey hunter knew that the bees sometimes swarmed in the cool interior of caves, so he climbed cautiously up to the largest of the Cub Caves and poked his long stick into the darkness, probing for honeycomb.

Unbeknown to the man, the young male leopard had entered the cave earlier that morning and now lay sleeping on the cool stone floor. Finding himself trapped by the man's silent approach, the young leopard lunged forward and sank his claws into the man's face. It was all over in seconds. As the man stumbled backwards with his hands clutching at his face, the leopard made good his escape. Just then, a carful of tourists from Governor's Camp arrived on the scene. Hearing what had happened, the driver took the man to the dispensary at Mara Rianta. Ever since that day, the young leopard had been even more wary of people moving about on foot, slipping away whenever he spied Masai heading in his direction.

It was extraordinary to see these two leopards together, strangers tolerating each other in a way contrary to our view of how the solitary cats are supposed to behave. The urge to mate was so strong that it dissipated their hunger for solitude, bringing them together for a few brief days in a union of extraordinary intensity. Their individual scent – distinguishable between males and females – fascinated them. They sniffed and sprayed, raked the ground and grimaced, caught up in an orgy of communication. But the male was too young and inexperienced to mate successfully. His presence simply frustrated the female, who at one point actually mounted him, grasping his neck in her jaws and thrusting impotently as she covered his body. Yet she could not abandon him.

Clouds merged to a uniform greyness, sweeping over the marsh towards Leopard Gorge. The heat died and the Paradise female began to stir. At four o'clock, she emerged from one of the caves and leaped into the Teclea bush screening the place where the male rested. The whole tree quivered as she sprang from branch to branch, descending again to groom in the long grass. She looked up to where the young male lay on his side, partially hidden along a ledge leading into a deep cave carved into the vertical rock face. It was one of his favourite hiding places. Often in the past he had wiled away the hours here, waiting for his mother to find him and lead him to a kill. Like all female leopards, the young male's mother had done everything possible to avoid other leopards, relying on scent messages and nocturnal calls to signal her presence in an area. So the young male was totally confused by the behaviour of the Paradise female. Their meeting had come as a surprise; it was she who had sought him out. Now she was greeting him in a manner that he had never experienced before.

That night the Paradise female killed a young impala and dragged it into one of the caves. Her ardour for the young male had finally waned, her desire for his company evaporating as her hunger increased. She did not share her kill with him and when he approached the place where she lay, she coughed and hissed, warning him to keep his distance. Her need for

Chui with a young impala kill. In areas where there are no large predators to steal their food, leopards have no need to carry their kill into a tree. Instead, they can safely feed in dense cover on the ground. Sixty per cent of an impala or gazelle is edible – the rest is discarded and the skeleton left for vultures to pick over. A leopard will often pluck the hair from the carcass as it feeds.

solitude had returned; sharing food with another is anathema to any leopard. Baboons heard the confrontation between the two predators and gave voice to their alarm. But the Paradise female ignored them. The night was her ally. Now, in the darkness, it was the baboons who were fearful. The troop spent the night in the fig trees at the western end of the gorge, sitting in twos and threes along the thick branches, huddled together for safety. Next morning the Paradise female slipped from the cave, abandoning the remains of her kill and distancing herself from the baboons.

Gradually the baboons left the trees, sitting in groups on top of the rocks, grooming and soaking up the warming rays of the sun. As they began to move along the lip of the gorge, one of the big males discovered the remains of the leopard's kill. He held the meat easily with his nimble fingers, tearing at the flesh with enormous yellow canines, intimidating other members of the troop who attempted to move closer. Every so often he glanced around him, as if nervous that the leopard might suddenly reappear and try to reclaim her kill. But by then

The Paradise female resting in a fig tree. Leopards often use fig trees as a comfortable and shady daytime resting place. There is surely no creature in the world that can relax quite like this.

the Paradise female was more than half a kilometre away on Fig Tree Ridge, resting in one of her favourite trees.

The young male tarried in the vicinity of Leopard Gorge for the next few days, hunting impalas and gazelles among the thickets. He was next seen wandering across Emarti ya Faru, heading south towards the Ngorbop lugga, where the old male leopard had been speared by the Masai. The young leopard was careful not to be seen by the Gorge pride whom he knew were somewhere in the vicinity. He had heard them roaring just before dawn. The male had learned how dangerous lions were from the day he was born, building on his instinctive fears by absorbing the wariness that his mother betrayed whenever lions were close at hand.

Perhaps it was the arrival of the two cars that prompted the young leopard to forsake the safety of his hiding place – just as the Gorge pride happened to be moving towards the place where he had been resting. If only he had seen the lions approaching he could simply have climbed one of the tall, slender trees growing out of the lugga, where no lion could follow.

ABOVE: *Lions are very aggressive towards other predators with whom they are competing for food or space, although they seldom feed from the carcass of another predator.*

The lions spotted the leopard hurrying away from the vehicles. One of the young male lions charged from close range. Others quickly joined in, biting and clawing at their opponent, who was easily outweighed by his rivals. They recognized him as an enemy and attacked him just as they would have attacked a lion from another pride. The leopard tried to defend himself, but to no avail; they were too many and too strong. Urged on by the visitors, the drivers of the cars attempted to separate the combatants, harassing the lions with their vehicles, their passengers outraged at what they perceived to be an unwarranted attack. But there was to be no last-minute reprieve. The leopard had been bitten in the neck and hindquarters, his abdomen torn open; he was doomed.

Next morning I drove to the spot where the incident had taken place. Vultures gathered in the treetops at the edge of the lugga. The leopard lay where he had died, his spotted face frozen in an ugly grimace. There could have been no more powerful a reminder of the antipathy that lions display to all the other predators, determined to protect their own interests against competitors for food and space.

The Bila Shaka pride, meanwhile, had vanished. They had not been seen since I had left them and headed for the gorge days earlier. This was a major blow for the drivers from Governor's Camp, as they were usually the easiest to locate of all the lion prides.

ABOVE: *The young male leopard from Fig Tree Ridge, killed by members of the Gorge pride.*
OVERLEAF: *These lionesses and their offspring spent an entire day alternately watching and harassing Halima, the female black rhinoceros.*

I continued towards Paradise Plain, skirting around the rocky bluff known as Rhino Ridge, whose scattered croton thickets were famous for harbouring a number of black rhinos in the 1970s. I can well remember embarking on a morning's game drive, knowing that I would see at least one or two rhinos. If I was lucky, I might find half a dozen, and on one particularly memorable afternoon I found nine rhinos between Musiara Marsh and Paradise Plain. But even then Kenya's rhinos were being heavily poached.

Today, Africa's rhino population is hurrying down the road to extinction, ruthlessly hunted for its horn. In the whole of Africa barely 3,500 have survived the poaching gangs. Thirty years ago there were more than that in Kenya's Tsavo National Park alone.

The last time I saw a rhino in the Kichwa Tembo area was in the early 1980s, and it was dead; a bloated, hornless, fly-infested carcass left to rot in the biting heat. Not long after that it became part of the safari routine for the drivers from Kichwa Tembo Camp to visit Rhino Ridge in search of a rhino named Halima. She was now the only one to be found anywhere in the northern Mara.

Everyone knew Halima. She was a gentle creature, perhaps too gentle. Her battles to rear her newborn calves amidst the lions of Rhino Ridge had become part of folklore. Halima's mother had been killed in 1978, during a period of intense poaching. Halima was two or three years old at the time – big enough to survive further attacks by lions. Like all rhinos, she was

In the 1960s, the Mara's black rhino population numbered more than a hundred. By 1983, only eleven had survived the onslaught of poachers. Today there are thirty-one rhinos in the Mara, thanks to the sustained efforts of the reserve authorities and Friends of Conservation who have provided vehicles and radio equipment to help with surveillance.

The skull is all that remains of a rhino killed by poachers. Rhino horn is much in demand in Northern Yemen, where it is carved into ornate handles for daggers, prized there as a symbol of manhood. Powdered rhino horn is also used in traditional medicine in the Far East. In fact, a rhino's horn is made of a compacted hair-like substance called keratin, which has no proven medicinal properties and, despite the widely held belief, no value as an aphrodisiac.

a creature of habit. She traversed a home range of approximately fifteen square kilometres, which provided her with sufficient browse, water, places in which to dust-bath and wallow, and a mate with whom to breed. Halima appeared a lonely figure without the great bulk of her mother by her side, standing out among the countless thousands of wildebeest during the dry season or marooned in a sea of grass during the rains. In 1981, she was shot in the shoulder by poachers, but recovered after being tranquillized and having the bullet removed. By 1983, only eleven rhinos remained in the Mara, out of a population that had numbered 108 in 1971.

Halima had produced three calves since 1981, the first of which was killed by the Ridge pride. She appeared strangely at a loss when the lions attacked her calf; she puffed and snorted, charging this way and that, scattering lions in all directions, but to no avail. There were just too many of them. Her second calf was again attacked by the lions, so it was decided to remove Halima and her calf to a safe haven on an ox-bow of the Mara River, not far from Governor's Camp. But the plan failed. The Marsh lions proved just as determined to kill the tiny calf as the Ridge pride had been. They came at night, badly mauling the baby rhino and tearing off part of one ear. At some point Halima decided to cross the river, leaving her calf stranded on the far bank. In their attempts to save the calf the rangers were forced to shoot some of the lions – highlighting the value that Kenya places on every last rhino.

Finally the calf was captured and flown to Nairobi, where he was cared for by Daphne Sheldrick until he was old enough to join the other sixty black rhinos living in Nairobi National Park. In 1985, another young rhino called Samia received considerable publicity when she was flown to the Mara, so that she might be blessed by Pope John Paul during the pontiff's visit to Kenya and raise awareness of their plight.

Halima soon became pregnant again and after a gestation period of fourteen months produced the third of her calves on 17 March 1988. Once more the rangers were forced to shoot a number of lions to protect it.

In April 1990, Halima was shot to death at point-blank range near Soit Naibor – the place of white stones – to the north-west of Mara Intrepids Camp. Ironically, the reserve authorities had been puzzling over how to ensure that Halima became pregnant again. Kioko, the sire of all of her previous calves, had crossed the Mara River in late 1988 or early 1989 to become the dominant bull in the Mara Triangle. Since then Kioko has mated with four different females, yielding a far greater return for his reproductive efforts in holding a territory than his years of lonely isolation monitoring Halima's movements.

Halima was the first rhino to be killed in the Mara since 1984 – an unqualified achievement when viewed against the enormous value of rhino horn to the poachers. This success has been brought about by the establishment of a specially equipped rhino surveillance unit, which monitors the rhinos on a daily basis. The population has more than doubled during the last six years and now stands at thirty-one animals. What was baffling and disturbing about Halima's death was that both her horns were found abandoned a few metres from her body. There, dumped unceremoniously on the ground, were five or six kilogrammes of horn, worth tens of thousands of dollars when cut up and carved into ornate dagger handles in Northern Yemen or ground into powder for medicinal purposes in the Far East. What possible reason could there be for killing a rhino if not for its horns? The only clue was

Some of the crocodiles in the Mara River weigh a thousand kilogrammes and measure over five metres in length. During the first river crossings of the dry season, the crocodiles are very active, taking every opportunity to prey on the herds as they cross. Even the cautious and orderly zebras are not immune.

in the occurrence of similar acts of wildlife vandalism elsewhere in Masailand. Over the years disgruntled Masai herdsmen have speared rhino and other wild animals in Amboseli National Park, whenever they have felt that their needs are not being taken seriously by the wildlife authorities. In effect they are saying, 'The priority should be people, not wildlife.' But whatever the reason, no one was in any doubt that Halima had been used as a pawn, sacrificed to further man's selfish interests. It was tragic that they should have picked on so harmless and so easy a target.

Some 300,000 visitors will journey to the Masai Mara during the coming year, representing a huge influx of foreign currency for Kenya. But it will be a safari without the familiar sight of Halima, who for fifteen years slumbered out on the open plains near Rhino Ridge, delighting visitors with her prehistoric presence.

The banks of the Mara River are constantly changing with the seasons. What had previously been an easy crossing could be a disaster this year for the returning herds, allowing the crocodiles to gorge themselves. It is said that a crocodile feasts so well on the proceeds of the river crossings that it barely needs to feed again until the migration returns months later.

I continued on my way to check the Paradise river-crossing sites: two much-favoured places where the wildebeest gather each year to ford the Mara River. Most of the animals had departed a week or so earlier; only the stragglers remained, determined to wait for the rains to commence before moving further south. The river boiled white and angry, hissing and rumbling on its tortuous route from the Mau Escarpment to Lake Victoria, via the Serengeti. A bull wildebeest lay exhausted on a ledge above the churning water. Every so often he glanced nervously about him, watching for the inevitable predator. The bull's right foreleg was bent awkwardly beneath him and each time he struggled to change position he revealed the bloody wound where his leg had broken clean through below the knee. It would be a long night. Eventually, the hyaenas or perhaps a leopard would smell the blood of the dying wildebeest and would draw closer to investigate. At least he was beyond range of the ripping lunge of the giant crocodiles that hunted for food along this stretch of the river.

A troop of baboons approached the water's edge, bending forward carefully to sip the

Over a three-day period, 700 wildebeest drowned at this well-used crossing site; in a bad year, when the Mara River is swollen and the banks are steep with few exit points, as many as 10,000 wildebeest may perish in the water. But disasters such as this have little impact on the overall size of the migratory population – their number is regulated by the availability of dry-season forage.

water. Crocodiles lay nearby, eyes closed. Daily they had taken up their positions, like sentries guarding the shallows, motionless as the herds drew closer. Some of the massive reptiles had fed so well that they hardly need bother eating again until the migration returned. The hiss and squawk of vultures, the lowing and grunting of the herds – all was now quiet. The bare branches of the giant strangler fig were empty of the griffon vultures that had perched like huge brown and white leaves, awaiting their chance to feed on the carcasses. The feast was virtually over for another year – the stench of death had faded. Dozens of skulls and fragments of bone lay like weathered pebbles on the bank, cracked open by the hyaenas for the marrow. Long fly-whisk tails lay half buried in the dried mud, scattered among the pearl-lined shells of freshwater molluscs. It looked like a hastily uncovered mass grave.

The rain-filled sky glowered darkly. Next year at the cul-de-sac it would be even more difficult for the animals trying to cross. The far bank of the river had fallen away under the weight of the herds, leaving a vertical cliff of mud, six metres high. The rains – when they came – would fashion new obstacles for the herds. But although thousands of wildebeest die in the river crossings each year, the number of casualties pales in comparison to the size of the migratory population.

I headed back towards Musiara Marsh, anxious to locate the Bila Shaka pride. By chance I crossed paths with the six lionesses from the Paradise pride, clustered together with their eleven cubs on top of a huge termite mound. They had drifted north, skirting the western end of Rhino Ridge while the Bila Shaka pride hunted elsewhere. Paradise Plain itself was covered with long dry grass, drained of its nutrients by the withering sun and avoided by the scattered herds of wildebeest and zebras. They had massed on the fire-scorched area to the east of Governor's Camp, where luxuriant green shoots were bursting through. Twice that morning the Paradise lionesses tried to ambush zebras. And twice the wind carried their scent to the animals well before they could get close enough to launch a successful attack. Without the dense herds of wildebeest daily plunging through the long grass towards them, all the lion prides were having to work harder to feed themselves and their cubs. There were few easy kills.

The following morning I found the Bila Shaka pride lying flanked around the spring feeding the marsh. The cool air formed a blanket of mist which lay over the predators; but before long they were forced to move on, steamrollered back towards cover by a party of old bull buffaloes. At midday they were on the move again, this time summarily dismissed from the forest edge by a herd of elephants. Watching from the treetops, another predator preened and stretched. Dressed in vivid black and brown plumage, with coral-pink feet and face, the bateleur eagle surveyed the wide spread of the marsh, keeping an eye out for any clue that would guide him to his prey: snakes and grass rats, carrion and termites. Low and fast, hugging the edge of the valley, came a martial eagle, a giant among Africa's birds of prey. The female martial sailed upwards, putting the bateleur to flight. Soon afterwards she was joined by her mate, who was dwarfed by her size. Together they preened and displayed, holding fast to their perch in the sky with massive yellow feet, bobbing their heads in breeding nuptials.

With the dry season refusing to release its stranglehold on the surrounding plains, the spring waters of Musiara Marsh beckoned to the Masai herdsmen. Soon their journey became a daily ritual. There were no crocodiles to worry about here. Each day during the hottest hours, they marched their cattle along the edge of the lugga, heading for the spring.

Forewarned by the tinkling of cowbells and the shrill whistles of the herdboys, the Bila Shaka lions peered through the bushes. It would have been easy for them to have crept into the open and stalked the livestock. Over the centuries man has bred most of the fierceness and speed of foot out of his domestic animals. The cows would have had no chance of defending themselves against a lion. But the Bila Shaka pride remembered the time that the Masai had retaliated for the death of a cow near the Musiara entrance gate, rushing at them with spears and killing one of the cubs.

At first the pride held their ground. But when another herd of cattle moved closer to the lions' resting place, the predators knew they must move from the marsh. They got to their feet and melted away through the trees. Brown led the way, weaving through the last of the long grass, heading for the river. For as long as they stayed within the reserve the lions were safe.

During the next few weeks the Bila Shaka lionesses and their cubs abandoned Musiara Marsh, spending most of their time hunting at the edge of the Kichwa Tembo pride territory

OVERLEAF: *The migratory herds must drink every day during the dry season, when there is little or no moisture left in the grass. Wildebeest and zebras often come to the river in the middle of the day when the sun is at its hottest. There are a quarter of a million zebras in the Serengeti-Mara ecosystem. The large migratory herds are made up of family units comprising a stallion, several females and their offspring.*

During the migration, lions often lie in ambush at river-crossing sites. In fact, this zebra escaped, as do the majority of prey that lions attempt to hunt. Zebras are not ruminants and have sharp upper and lower incisors for chopping down the coarse grass stems and seed-heads that are best suited to their digestive systems.

on the other side of the river. With only the Old One's two daughters occupying the area, they could afford to trespass with impunity. Meanwhile, the two males from the Bila Shaka pride seemed quite content to stay where they were. Already, one of the Kichwa Tembo lionesses had given birth to four cubs sired by one of the males.

Finally, after days of waiting for the heat to dissipate, it rained as I sat watching the Bila Shaka pride near Kichwa Tembo Camp. I was overjoyed. At last I could breathe again, refreshed by the chill of the air. The force of the storm rocked my car on its springs, great waves of spray gusting in billows across the grass as the rain came lashing down. The Bila Shaka lions bowed their heads, squatting flat against the ground, trying to shelter behind clumps of long grass. Only the black stripes across the backs of their ears stood out, marking their presence. Hail cracked against the windows and beat on the roof, raining down on the lions, knocking leaves from the trees. The intensity of the storm was too much for the younger pride members. They rose to their feet and fled. They had never seen anything like this before. Across the plain a thin line of trees stood out like skeletons draped against the grey blanket of falling rain; but there was no shelter there. Nothing could lessen the stinging pain of the hailstones. The lions ran this way and that like scalded cats, ears back, tails hung low. Half an hour later the storm passed on, ending as suddenly as it had begun, sweeping south to moisten the exhausted grasslands on Paradise Plain.

In a matter of days, the rains rejuvenate the parched grasslands. The herbivores – even the normally alert topi – are less vigilant when it rains – a fact that does not pass unnoticed by the predators. The short rains herald the arrival of the young of many prey species – topi, warthogs and gazelles.

After the storm, the silence was absolute, as if the land had been battered senseless. Then birds began to call and the drum-rolls of thunder echoed back across the plain. The grass glistened with droplets of rain. One of the lionesses stood up and rubbed her head against another, then flopped down in front of her, inviting her sister to groom the rain from her back. Soon the lions began to look for prey.

I struggled to count them among the tangle of bushes in the floor of the Miti Mbili lugga. Two of the four lionesses were missing, but still there was no doubting who these lions were. Grey Coat, with her distinctive pelage and flop ear, identified them as the Bila Shaka pride. Their absence from the Miti Mbili lugga had lasted the better part of a year.

Time spent outside her pride territory had cost Grey Coat dearly. She walked now with a barely disguised limp and her spine bore a number of angry swellings that protruded from the sway of her back. Her coat was torn and patchy. The punctures were still clearly visible, and the surrounding skin was licked bare, where she had cleaned the wounds. The old lioness had undoubtedly been in a fight with other lions. She was beginning to remind me of the Old One.

Epilogue

Man's fundamental desire is to be God.

JEAN PAUL SARTRE

By the vanity of the same imagination he equals himself to God, attributes to himself divine faculties, and withdraws and separates himself from all other creatures; he allots to these, his fellow companions, the portion of faculties and power which he himself thinks fit.

THE DEFENCE OF RAYMOND SEBOND
MONTAIGNE

THE GREEN OF THE RAIN-SOAKED PLAIN stuns the senses. It is as if the whole world were alive with animals. Gazelles in their tens of thousands speckle the earth, topi stand like statues, ever vigilant, nursing calves – still wet – on wobbly legs. Parties of warthogs with piglets – miniature replicas of the adults – nuzzle the fresh shoots on bended knees.

The richness of the scene dazzles the mind, the piercing sun tempered by the breeze. It is so bright that reflections play hard on the eyes. Zebra stallions fight for the right to breed with females that only a week ago gave birth to foals; white blossoms abound like confetti tossed carelessly across the plain. At the edge of the Miti Mbili lugga a herd of buffalo – massive bovine shapes – plough a dark furrow through the rolling sweep of the plain – watched by a hyaena poised among the shadows, dark eyes fixed on a calf struggling to keep up with the herd.

As I jot down my thoughts, I realize that nothing has changed. The scene I am describing is little different from the one I witnessed fifteen years ago. The endless cycle of life and death is everywhere repeated. There is a sense of times past, of animals existing as they did long before man appeared on the savannahs of Africa.

OPPOSITE: *Yellow-billed and red-billed oxpeckers feed primarily on ticks and biting flies, but sometimes peck flesh from open wounds on grass-eating animals. They also warn the host animal of danger by flying up and alarm-calling when disturbed. Although they associate particularly with buffaloes, rhinos and giraffes, they are also seen on impalas, hippos and zebras. This zebra has obviously survived a lion attack which has left its rump badly scarred.*

ABOVE: *Cheetahs fare well among the Masai pastoralists because there are fewer lions and hyaenas around to compete for food or kill their cubs. Both these predators are known to kill livestock, prompting the Masai to protect their interests with spears and poison. Cheetahs do not scavenge and therefore do not succumb to poison; nor do they take livestock except in areas where natural prey populations have been depleted by man.*

RIGHT: *The African night is filled with the vibrant chant of frogs. Servals, birds of prey, snakes and even leopards will kill them if there is no more substantial prey available, but the pattern of this tree frog's skin camouflages it perfectly amid the cow dung in an abandoned Masai manyatta. During the dry season, some frogs aestivate – the hot weather equivalent of hibernating – concealing themselves in moist burrows to avoid desiccation.*

I leave my vehicle and walk through the narrow entrance of one of the Masai's temporary thorn-bush enclosures, long since deserted. Three makeshift shelters snuggle against the circle of thorns, each one an igloo of branches thatched with grass and cow dung, supported by a single main timber set astride forked centre posts. The branches are secured with strips of bark, twisted into knots to hold the structure firmly in place. Everything is crafted from nature. The ash of countless fires stands in a windswept pile in front of the simple shelter where the *morani* kept a look-out for predators and cattle rustlers. A day or so earlier I had watched as a cheetah and her twin yearling cubs explored the abandoned dwellings. Effortlessly they leaped on to one of the flat roofs, where they lay together, taking in the view of the surrounding countryside and searching for prey. Now a crombec is busy nesting beneath the arch of one of the shelters; its tiny nest is as carefully woven as the roof it hangs from. The broad leaves of a sodom apple plant erupt from the floor, where a frog sits so perfectly camouflaged among the cracked and weathered dung that I almost tread on it.

This is man at his least destructive, catering to simple needs, without extravagance, shorn of the wastefulness that is such a corrosive feature of Western society. Once abandoned, bomas such as these gradually blend with earth and wind. The long grass reaches up and joins hands with the thorn-bush; the cow-dung floor enriches the soil. In a year or so it will have been transformed into a dense patch of croton bush, seeded by the Masai cattle, providing shelter for lions and leopards, as well as food for elephants and rhinos.

Elsewhere, conical-roofed huts and corrugated-iron houses have the look of a tribe adapting to a more sedentary existence. Strips of plastic and empty containers are left scattered across the plains. Trees are hacked down for fuel; the threat of agriculture is never far away. The trickle of emigrants from the densely populated Lake Victoria Basin to the west is turning into a flood. Land disputes are everywhere gathering pace as Kenyans wrestle with the consequences of rampant population growth. Soon the last of the nomadic pastoralists will mark out their individual plots and fence them off, before anyone else moves on to their land.

The increase in non-Masai people – mainly Luos, Kipsigis, Nandies and Kisiis – has led to an upsurge in meat poaching on the periphery of the reserve. Currently the level of off-take by the poachers is sustainable, due to the large populations of wandering herbivores. But it cannot last forever. In response to this threat the Stock Theft Unit mounted a successful anti-poaching campaign along the Isuria Escarpment, leading to a decline in the number of animals trapped in wire snares. And a major training programme has just been completed by all reserve rangers. Fortunately, poaching of rhinos and elephants is virtually non-existent in the Mara. The vigilance of the reserve authorities and the presence of large numbers of tour vehicles make it difficult for poachers to operate undetected.

Ultimately the survival of the Mara and its wild animals depends on the goodwill of the local people – coercion won't work. To help achieve this, a community-based conservation education extension programme has been initiated by Friends of Conservation, with a grant from the Overseas Development Agency. The object is to involve the local community in the conservation of wildlife and other natural resources, while helping them support their pastoral lifestyle and meet their development needs.

OVERLEAF: *Many young Masai now attend school rather than spending their days looking after their family's herds. The Masai are being encouraged to abandon their nomadic lifestyle and many of their traditions, but colourful ceremonies remain an important part of their culture.*

Hunger and disease are the norm for many people in Africa. At present the herdsmen are understandably more concerned with their livelihood than with understanding what motivates a lion or an elephant. The wildlife sanctuary that provides tangible material profit is going to be the one that survives. For as long as the Masai see a financial return for the hundreds of thousands of visitors who drive annually across their land, there is hope. Only by removing poverty can the wilderness be saved.

The Northern Ranchlands are trust lands administered by the Narok County Council; they are potentially Africa's finest communally owned game ranch. At present the Masai living in the group ranches bordering the reserve receive twenty per cent of the daily fee collected from visitors to the Mara – a sum of five US dollars per visitor. In return for this revenue they must abide by a comprehensive land-use policy, which excludes subdivision or agriculture. The County Council intends introducing reserve bylaws further to protect the wildlife in the dispersal areas.

Wildlife-based tourism is Kenya's primary source of foreign exchange, producing US$480 million in 1990 – forty-three per cent of the country's total earnings. The Mara alone generates nearly two million dollars for the Narok County Council. To cope with the impact of all the tourist vehicles, rangers patrol game-viewing areas to try and prevent people harassing the animals, particularly cheetahs who need to secure their prey during the daytime, while their major competitors are lying up. The Mara is the only reserve in Kenya where off-the-road driving is allowed, which at times leads to predators being encircled by tour vehicles. A game-viewing plan, based on the construction and maintenance of a comprehensive network of all-weather roads, has already been outlined. When this project is eventually completed, it will no longer be permitted to drive off the roads.

'Sustainable utilization' has become the byword for many environmentalists. In southern Africa a blend of tourism, trophy hunting and game ranching for meat and wildlife curios helps to fund conservation and provide revenue for people living in wildlife areas. But a purely materialistic approach is not enough to save wilderness. Profits from tourist-based wildlife conservation can always dry up, as the Gulf War so clearly proved.

There is nothing new about exploiting natural resources for economic gain. Forty years ago Aldo Leopold, the American conservationist, pointed out its limitations:

> A system of conservation based solely on economic self-interest is hopelessly lopsided. It tends to ignore, and thus eventually to eliminate, many elements in the land community that lack commercial value, but that are (as far as we know) essential to its healthy functioning.

Surely idealism has a part to play in conservation. As HRH the Duke of Edinburgh, president of WWF International, said:

> The need for someone to stand up and champion nature, and speak for the earth with wisdom and insight, is urgent.

I decide to take one last game drive before heading for Nairobi. First I track down the Bila Shaka pride, and find them slumbering in the shade of the Miti Mbili lugga. For a while during 1990 they had been Marsh lions again. Now they are back at Miti Mbili. Within the next few months the nine cubs will become semi-independent, old enough to prompt their mothers to come into season. The two young Serengeti nomads that were received

A sub-adult male and female of the Bila Shaka pride play at Musiara Marsh. Pride members seem to enjoy each other's company and cubs of all ages play in the cool of dawn and dusk. Play helps reaffirm social bonds among members of a pride and allows cubs to strengthen their muscles and refine their hunting skills.

with such aggression by Grey Coat and her relatives are still moving throughout the Musiara area. Perhaps it will be them that father the lionesses' cubs: an inevitable transfer of power from one coalition of males to another. The four male cubs will soon be ousted from the pride. But what about their five sisters? Will they be recruited into the Bila Shaka pride? Or – just as happened ten years ago – will the pride territory be divided among female relatives?

So much has happened from the perspective of the individual animals – or so it seems. But the truth is – despite the many hours of detailed observations – we know little about predators. Deductions made from even the most scientific of studies represent good guesses, acknowledging how incomplete is our sense of worlds outside our own. Generalizations mask the uniqueness of the individual, clouding our perceptions; we cease to let them be themselves, imprisoning them by our superficial descriptions. Biologists working with the twenty-five-year-old Serengeti lion project have given us a new lion, one separated from

The Masai have always been tolerant of wildlife and are rarely involved in the illegal slaughter of animals for profit. Elephants thrive outside the Mara Reserve and as long as the Masai remain a pastoral people, there is little cause for conflict. But in areas where maize and other food crops are grown, elephants soon become a menace.

folklore. But they have not found the whole truth; the animals' inner lives remain a mystery. As Barry Holstun Lopez wrote in *Of Wolves And Men*:

> I remember . . . recalling Joseph Campbell, who wrote in the conclusion to *Primitive Mythology* that men do not discover their gods, they create them. So do they also, I thought . . . create their animals.

I head north, indulging the urge to drive through Leopard Gorge. The cluster of vehicles gathered in the bottom of the gorge pinpoints the place where the Paradise female has chosen to rest. She lies there in the leafy crown of a fig tree, her head turned away, ignoring the outbursts of laughter, the chatter of drivers. Finally, when she leaves her aerial perch and slips away, everyone falls silent. She creeps along the rocky horizon, pausing only to stare towards the place where two lionesses lounge on their sides, each with a litter of young cubs clinging to her teats, the cubs' bellies swollen with milk. The leopard knows they are there, and is gone, leaving unanswered the question of her pregnancy. The search for answers, which gives meaning and tension to the watching, is what draws me back again and again to this place.

The transition from wilderness gathers pace as I negotiate a way through the rutted track

The Masai are increasingly being called upon to participate in Kenya's cash economy. It is essential that they continue to receive adequate compensation for allowing wildlife to share their valuable land. For as long as the Masai are allowed to preserve their traditional way of life, there is hope for the Mara's incomparable wild places.

passing Aitong Hill, the haunting cries of the wild dogs just a memory. Facts and figures pass before my eyes – the grim reality of our future imposed on the beauty of nature. Tourism will be the world's biggest industry by the year 2000 – but at what cost to the environment? The human population will double in fifty years to ten billion – Africa's population of 630 million will soar to one billion by 2000. By then one million kinds of animals and plants will be extinct. Yet biological diversity is the key to the earth's ecosystem, helping to regulate the climate and generate soils, providing food and medicines as yet unknown to man.

A sign alongside one of the Masai bomas announces the Ololturoto Cultural Manyatta: 'Price of entry one hundred Kenya shillings, no sweets or biscuits to children.' Fifteen schoolboys hurry through acacia thickets that at times provide shade for buffaloes and lions. The boys are clad in blue shirts and red shorts, school books tucked inside cowhide satchels slung over their shoulders. Some wear traditional red *shukas* over their school uniforms, the wind revealing the changes. In years to come they will understand why it is important to cherish wildlife and to sell cattle on a sustainable basis rather than watch the inevitable die-off in times of drought. They will be proud that their land harbours one of the last great wildlife sanctuaries. And they will be able to enjoy the sight of wild lions – hear their roars echoing across the spotted landscape as they recall the part these magnificent creatures have played in moulding their colourful traditions in the Kingdom of Lions.

Acknowledgements

I wish to express my gratitude to the government of Kenya for granting me permission to live and work in the Masai Mara, and to the Narok County Council who administer the Masai Mara and were generous in their support of this project.

Senior Warden James Sindiyo has been extremely helpful with his advice and co-operation over the years. Dr Holly Dublin (WWF) and Dr Pieter Kat (Genetics Department, Nairobi Museum) patiently listened to my questions on matters of science. Joseph Ogutu, a graduate assistant from Moi University, generously shared the findings of his MSc lion project in the Mara.

Jock Anderson, the managing director of East African Wildlife Safaris, has continued to be of great support, allowing me to use his office in Nairobi, where Sharon Fox and Stephen Masika keep me supplied with mail, messages and vehicle spares.

I would like to thank the Turner family, who run East African Ornithological Safaris, and Micheo Gitonga, the manager of Mara River Camp, who was helpful and hospitable whenever I visited my old home.

Since 1981 Geoff and Jorie Kent (of Abercrombie and Kent) have provided me with a base at Kichwa Tembo Camp, and enabled me to travel to fund-raising events in the US and UK on behalf of Friends of Conservation. Sammy Mwaura and David Markham at Abercrombie and Kent in Nairobi, and Martin Thompson (UK) and Alistair Ballantine (USA) have always been supportive of whatever project I happen to be working on.

Anne Taylor-Kent generously let me stay in her house at Kichwa Tembo overlooking the beautiful Sabaringo lugga. Maurice and Monica Anami, Charles Davis and James Mahaini accommodated my many requests for help in their capacity as managers of the camp. In the vehicle workshop, Anunda tended my aged Toyota Landcruiser, conjuring all kinds of miracles to keep me on the road. It is not just the luxury of being able to soothe a tired back in a swimming pool blending perfectly with a sweeping view of the Mara plains. Nor is it the undoubted quality of the food. It is the friendliness and excellence of the staff that makes living at Kichwa Tembo such a pleasure.

The drivers at Kichwa Tembo helped track the individual lions, leopards, cheetahs and wild dogs that I was following, as did many of the drivers at Governor's Camp, where Stephen Mutua was particularly helpful.

At Windsor Hotels (who manage Kichwa Tembo) David Stogdale, Peter Ngori, Guy Epsom, Paul Kier and many other members of staff were extremely helpful in ensuring efficient communication between Kichwa Tembo and Nairobi. Windsor also provided funding for the launch of my book on wild dogs at the Nairobi Musuem in May 1991.

John Buckley, the managing director of Air Kenya Aviation, enabled me to fly standby on their scheduled service to and from the Mara. A forty-five minute flight is heaven compared to a five-hour drive by road!

Mehmood Quraishy at Colour Centre helped to correct the photographic proofs of *Kingdom of Lions* and has always been generous with his time and advice on photography.

As in the past, Boris Tisminieszky shared his expertise with computers, rescuing me on a number of occasions when all seemed lost – literally! When 'Fatal Error' appears on the screen of your laptop, you need friends like Boris.

Just before completing this book I attended the ninety-seventh birthday party of my old friend Colonel T. S. Conner, DSO, KPM. For many years the colonel made his house in Nairobi a home for me whenever I was on leave from life in the bush. He continues to serve the same inimitable combination of good company and delicious curry to my family whenever they visit Kenya. The colonel is the kind of friend that all of us would like to be. I wish him continued health and happiness as he approaches his century.

In America, Greg and Mary Beth Dimijian sent me the latest scientific papers, allowing me to keep abreast of research into animal behaviour. Over the years, David Goodnow has provided a blend of good humour, camera equipment, Kodachrome 64 and sound advice. In England, Pippa Millard, Gillian Lythgoe, Jennifer Jeffries, Fran Norgren and Carla Stubbs were – as always – cheerful and hospitable despite my many demands on their time.

Dr Michael and Sue Budden have been the most marvellous friends ever since our first meeting at Kichwa Tembo in 1984. Nothing has been too much trouble and no request has gone unanswered, whether it entailed sending computers and printers for repair, shopping for car spares or ferrying me to and from Terminal 4 at Heathrow. Michael and Sue are avid conservationists and keen photographers, and were able to enjoy the unique experience of a safari through Kenya's wild places with my old friend Joseph Rotich before his death in 1990. Joseph inspired me with his great dignity. I still miss his warmth and humour.

I feel very fortunate to have been published by Kyle Cathie, now of Kyle Cathie Ltd. We first worked together in 1984 when Kyle was at Elm Tree Books and I was producing *The Leopard's Tale*. It is not often that one finds a publisher who is able to turn an author's dream into reality. It just goes to prove what a small team of real professionals can achieve in their back garden!

Caroline Taggart has co-ordinated and edited all my books, firstly while she was with Elm Tree and now as a freelance editor. Caroline has that essential quality as an editor of being able to keep morale high whenever the 'words' get tough, while bringing out the best in one's writing. Her own love of Africa has contributed enormously to these projects.

Brian Jackman, who was for many years the wildlife correspondent with the *Sunday Times* in London and is now a freelance travel and conservation writer, worked wonders with an early version of my text. His writing skills were amply demonstrated when we worked together to produce *The Marsh Lions*. Brian's evocative style has continued to inspire me since I started to write my own books.

It has been a tremendous pleasure to work with a designer of the calibre of David Fordham, who was prepared to listen to my views and blend them with his own creative skills.

Mike Shaw, my literary agent at Curtis Brown, and his assistant

Marion Cookson have weathered all manner of storms on my behalf. If you are thinking of writing a book, get an agent like Mike!

In Nairobi, my wife Angela accomplished much more than I asked of her, all without fuss. Her sense of design and artistry, and love of the bush, made my life even more fulfilling.

My family in England have been of enormous support to me, encouraging me in every possible way to pursue the career of my choice. I could have asked for nothing more.

Vehicle

I have always driven a Toyota Landcruiser, preferring its spacious interior and rugged reliability to that of other four-wheel-drive vehicles. Jan Thoenes at Toyota Kenya, in conjunction with Toyota Japan, generously agreed to lend me one of their new five-speed Toyota Landcruisers. This enabled me to 'retire' my twelve-year-old vehicle, which had given me years of good service in the Mara and Serengeti.

Mike Carnall, Mr S. Shah, Rashid Yusuf and the staff at Toyota's industrial workshop customized a body for the new Landcruiser to accommodate the nomadic existence of a wildlife photographer – including a comfortable bed (which doubles as storage space for food and vehicle spares), fuel-carrying capacity equivalent to 2,000 kilometres, and a reliable winch.

Photography

My first 35 mm camera was a Canon EF, purchased to record my overland trip through Africa in 1974. The EF was one of the first models offering automatic functions featuring shutter priority. Only later did I discover that Canon lead the field in developing the 'fast' telephoto lenses that are so essential for wildlife photography. Now that autofocus technology has reached the point where it offers real advantages for professional photography, Canon have once again proved to be the best.

I use three Canon T90 camera bodies and one Canon EOS1 autofocus camera, all of which feature highly reliable metering systems. I cannot imagine undertaking a photographic assignment without carrying two much-favoured zoom lenses – a Canon EF 20–35 mm f2.8 and an EF 80–200 mm f2.8 (autofocus). Of my longer telephoto lenses, the 300 mm f2.8, 500 mm f4.5 and 800 mm f5.6 have all provided years of reliable photography, despite being dropped on a number of occasions! Another piece of equipment that I consider essential is a × 1.4 extender. This is a compact, high-quality optical converter that extends the magnification of a lens while only incurring the loss of a single stop of light; the minimum focusing distance of the prime lens remains the same. Carrying an extender helps safeguard against the sudden loss of a telephoto lens while on assignment.

For the past ten years Chris Elworthy, marketing manager at Canon (UK) Camera Division, has kept me up to date with details of the latest range of cameras and lenses. Chris has enabled me to have my equipment cleaned and serviced at a day's notice whenever visiting England; dust and vibration make regular servicing essential. But be sure to let Canon know well ahead of time if you need their expertise. In this respect Lynne Scott at Canon's service department has been particularly helpful.

The majority of the pictures in this book were taken on Kodachrome 64 film, processed in Switzerland. I also used Fuji Velvia 50 (exposed at ASA 40), processed at Ceta Laboratories in London.

Drawings

All the drawings were produced in pen and ink using Rotring pens (0.1 and 0.2 nibs). Some of them are now in private collections, so I would like to thank my mother Margaret Scott (lioness and cub), sister Caroline Scott (Thomson's gazelle and fawn), Sue and Ed Tweddell (waterbuck), Colonel T. S. Conner (buffalo), Bob Mottram (male lion – Old Man, and impala with calf) and Bill Wood (lioness carrying cub, and elephants), who own the originals of my work.

Anyone interested in finding out more about the Mara or contributing to its future may like to contact:

Friends of Conservation, 1520 Kensington Road, Oak Brook, Il 60521–2016, USA

or

Sloane Square House, Holbein Place, London, SW1W 8NS, England.

Bibliography

It is not appropriate in a book of this kind to cite individual references throughout the text. Instead, this bibliography contains popular and scientific accounts of predator biology which have proved particularly helpful in expanding on my own observations in the field. I was also fortunate in being able to enjoy stimulating conversations with Dr Holly Dublin (on the ecology of the Mara), Ben Kipeno (on the Masai, and FOC's Community Conservation Programme), Joseph Ogutu (on his lion research in the Mara), and Judith Rudnai (on her four-year study of lions in Nairobi National Park, from 1968–72).

However, they and the authors listed below remain blameless for any inaccuracies in my text, or for the inevitable simplifications that I have made in interpreting their work.

Ames, E. 1968. *A glimpse of eden*, London: Collins

Bertram, B. C. R. 1978. *Pride of lions*, London: J. M. Dent

Broad, S. 1991. *Contracting cats, BBC Wildlife*, Vol. 9, No. 3, p. 201

Cooper, S. M. 1989 (May). *Rival predators gang up against each other, Johannesburg Sunday Times Magazine, RSA*

– – 1991. *Optimal hunting group size: the need for lions to defend their kills against loss to spotted hyaenas. Afr. J. Ecol.* 29: 130–136

Dublin, H. T. 1984. *The Serengeti-Mara ecosystem, Swara*, Vol. 7, No. 4

– – 1986. *Decline of Mara woodlands: the role of fire and elephants*, University of British Columbia, D.Phil. thesis

Dublin, H. T., and Chara, P. S. 1991. *A report on some aspects of ecological monitoring in the Masai Mara National Reserve and adjacent pastoral lands*

Fanshawe, J. H., Frame, L. H., and Ginsberg, J. R. 1991. *The wild dog – Africa's vanishing carnivore, Oryx*, Vol. 25, No. 3

Gellhorn, M. 1983. *Travels with myself and another*, New York: Eland Books, London & Hippocrene Books, Inc

Gilbert, D. A., Packer, C., Pusey, A. E., and O'Brien, S. J. 1991. *Analytical DNA fingerprinting in lions: parentage, genetic diversity and kinship, Journal of Heredity*

Ginsberg, J. R. 1991. *Studying the 'painted wolf' in Hwange, Zimbabwe Wildlife*, No. 62

Hanby, J. P., and Bygott, J. D. 1983. *Lions' share: the story of a Serengeti pride*, London: Collins

Harcourt, A. H., Pennington, H., and Weber, A. W. 1986. *Public attitudes to wildlife and conservation in the Third World, Oryx*, Vol. XX, July 1986. Oxford: Blackwell Scientific Publications Ltd

Jackman, B. J. & Scott, J. P. 1982. *The marsh lions*, London: Elm Tree Books

Kingdon, J. 1971–1982. *East African mammals: an atlas of evolution in Africa*, Vols. 1–7, New York: Academic Press

Kruuk, H. 1972. *The spotted hyaena*, Chicago: University of Chicago Press

Laurenson, K. 1991. *Cheetahs never win, BBC Wildlife*, Vol. 9, No. 2

Lavigne, D. 1991. *Human nature: slipping into the market place, BBC Wildlife*, Vol. 9, No. 2

– – 1991. *Human nature: your money or your genotype, BBC Wildlife*, Vol. 9, No. 3, pp. 204–5

Leopold, A. 1987. *A sand county almanac: and sketches here and there*, Oxford: University Press

Lopez, B. H. 1978. *Of wolves and men*, New York: Charles Scribner's Sons

Mabey, R. 1991. *Filthy images, BBC Wildlife*, Vol. 9, No. 3, p. 153

Moss, C. 1975. *Portraits in the wild*, Chicago: University of Chicago Press

O'Brien S. J., Wildt, D. E., and Bush, M. 1986 (May). *The cheetah in genetic peril, Scientific American*, pp. 84–92

Owens, D. D. & Owens, M. J. 1985. *Cry of the Kalahari*, London: Collins

Packer, C., and Pusey, A. E. 1982. *Co-operation and competition within coalitions of male lions: kin selection or game theory? Nature* (London), 296:740–742

– – 1983. *Adaptations of female lions to infanticide by incoming males, American Naturalist*, 121:716–728

Packer, C., and Pusey, A. E., Rowley, H., Gilbert, D. A., Martenson, J. S., and O'Brien, S. J. 1991. *Case study of a population bottleneck: lions of the Ngorongoro Crater, Conservation Biology*, 5:1–12

Packer, C., Scheel, D., and Pusey, A. E. 1990. *Why lions form groups: food is not enough, American Naturalist*, Vol. 136, No. 1

Rudnai, J. 1973. *Reproductive biology of lions (Panthera leo massaica Neumann) in Nairobi National Park, East African Wildlife Journal*, 11:243–251

– – *The social life of the lion: a study of the behaviour of wild lions (Panthera leo massaica [Neumann]) in the Nairobi National Park, Kenya*, Lancaster, England: Medical and Technical Publishing Co Ltd

Schaller, G. B. 1972. *The Serengeti lion: a study of predator-prey relations*, Chicago: University of Chicago Press

– – 1973. *Serengeti: a kingdom of predators*, London: Collins

Scott, J. P. 1980. *Wild dogs: the sociable predators, Swara*, Vol. 3, No. 1 pp. 8–11

– – 1985. *The leopard's tale*, London: Elm Tree Books

– – 1988. *The great migration*, London: Elm Tree Books

– – 1991. *Painted wolves: wild dogs of the Serengeti-Mara*, London: Hamish Hamilton

Seidensticker, J., and Lumpkin, S. 1991. *Great cats*, London: Merehurst Ltd

Sinclair, A. R. E., and Norton-Griffiths, M., (eds.). 1979. *Serengeti: dynamics of an ecosystem*. Chicago: University of Chicago Press